Brassey's
Battlefield Weapons Systems
& Technology, Volume VI

COMMAND CONTROL AND COMMUNICATIONS (C^3)

Brassey's Battlefield Weapons Systems and Technology Series

General Editor: Colonel R G Lee OBE, Formerly Military Director of Studies at the Royal Military College of Science, Shrivenham, UK

This new series of course manuals is written by senior lecturing staff at RMCS, Shrivenham, one of the world's foremost institutions for military science and its application. It provides a clear and concise survey of the complex systems spectrum of modern ground warfare for officers-in-training and volunteer reserves throughout the English-speaking world.

Introduction to Battlefield Weapons Systems and Technology—R G Lee

Volume I	Vehicles and Bridging—I F B Tytler *et al*
Volume II	Guns, Mortars and Rockets—J W Ryan
Volume III	Ammunition (Including Grenades and Mines)—K J W Goad and D H J Halsey
Volume IV	Nuclear Warfare—L W McNaught
Volume V	Small Arms and Cannons—C J Marchant Smith and P R Haslam
Volume VI	Command, Control and Communications—A M Willcox, M G Slade and P A Ramsdale
Volume VII	Surveillance and Target Acquisition— A. L. Rodgers *et al.*
Volume VIII	Guided Weapons (Including Light, Unguided Anti-Tank Weapons) — R G Lee *et al.*
Volume IX	Military Data Processing and Microcomputers—J W D Ward and G N Turner
Volume X	Military Ballistics—A Basic Manual—C L Farrar and D W Leeming
Volume XI	Military Helicopters—P G Harrison and J Everett Heath

For full details of these titles in the series, please contact your local Brassey's/Pergamon office

Other Titles of Interest from Brassey's Defence Publishers

CLOSE	Time for Action
RUSI	Nuclear Attack: Civil Defence
RUSI	RUSI/Brassey's Defence Yearbook 1984
SIMPKIN	Human Factors in Mechanized Warfare
SORRELS	US Cruise Missile Programs: Development, Deployment and Implications for Arms Control

COMMAND CONTROL AND COMMUNICATIONS (C^3)

A. M. Willcox,
M. G. Slade and P. A. Ramsdale
Royal Military College of Science, Shrivenham, UK

BRASSEY'S DEFENCE PUBLISHERS
PENTAGON–BRASSEY'S INTERNATIONAL DEFENSE PUBLISHERS
members of the Pergamon Group

OXFORD · WASHINGTON D.C. · NEW YORK
TORONTO · SYDNEY · PARIS · FRANKFURT

U.K	Brassey's Publishers Ltd., a member of the Pergamon Group Headington Hill Hall, Oxford OX3 0BW, England
U.S.A.	Pentagon-Brassey's International Defense Publishers, 1340 Old Chain Bridge Road, McLean, Virginia 22101, U.S.A.
	Pergamon Press Inc., Maxwell House, Fairview Park, Elmsford, New York 10523, U.S.A.
CANADA	Pergamon Press Canada Ltd., Suite 104, 150 Consumers Road, Willowdale, Ontario M2J 1P9, Canada
AUSTRALIA	Pergamon Press (Aust.) Pty. Ltd., P.O. Box 544, Potts Point, N.S.W. 2011, Australia
FRANCE	Pergamon Press SARL, 24 rue des Ecoles, 75240 Paris, Cedex 05, France
FEDERAL REPUBLIC OF GERMANY	Pergamon Press GmbH, Hammerweg 6, D-6242 Kronberg-Taunus, Federal Republic of Germany

First edition 1983

Library of Congress Cataloging in Publication Data
Willcox, A. M.
Command, control & communications (C3).
Includes index.
1. Command and control sytems—Handbooks, manuals, etc.
I. Slade, M. G. II. Ramsdale, P. A. III Title.
IV. Title: Command, control, and communications (C3)
UB212.W54 1983 335.3'3041 83-11844

British Library Cataloguing in Publication Data
Willcox, A. M.
Command, control and communications (C3).
—(Brassey's battlefield weapons systems and technology series)
1. Communications, Military
I. Title II. Slade, M. G.
III. Ramsdale, P.A.
623.7'34 UA940
ISBN 0-08-028332-2 (Hardcover)
ISBN 0-08-028333-0 (Flexicover)

In order to make this volume available as economically and as rapidly as possible the author's typescript has been reproduced in its original form. This method unfortunately has its typographical limitations but it is hoped that they in no way distract the reader.

The views expressed in the book are those of the authors and not necessarily those of the Ministry of Defence of the United Kingdom.

Printed in Great Britain by A. Wheaton & Co. Ltd., Exeter

Preface

The Series

This series of books is written for those who wish to improve their knowledge of military weapons and equipment. It is equally relevant to professional soldiers, those involved in developing or producing military weapons or indeed anyone interested in the art of modern warfare.

All the texts are written in a way which assumes no mathematical knowledge and no more technical depth than would be gleaned from school days. It is intended that the books should be of particular interest to army officers who are studying for promotion examinations, furthering their knowledge at specialist arms schools or attending command and staff schools.

The authors of the books are all members of the staff of the Royal Military College of Science, Shrivenham, which is comprised of a unique blend of academic and military experts. They are not only leaders in the technology of their subjects, but are aware of what the military practitioner needs to know. It is difficult to imagine any group of persons more fitted to write about the application of technology to the battlefield.

This Volume

This book introduces the fundamentals and theory behind military command, control and communications (C^3) systems and is intended for those who wish to widen their professional knowledge. Both current and future communications technology is examined to allow the reader to appreciate its vital support of command and control. The problems of communicating in an era becoming more and more affected by electronic warfare, are described. Due to the vital role played by communications on the battlefield a good understanding is essential for those who are to use C^3 systems.

Shrivenham. March 1983

Geoffrey Lee

Acknowledgements

A strength of the Royal Military College of Science is its unique mixture of military and academic staff. In writing this book we have been able to blend our diverse academic, industrial and military experiences.

Communications engineering is a highly mathematical and technical area and this can often disguise the basic principles from the non-specialist. We have attempted to reveal these principles and their relevance to military command and control by giving explanations in simple terms.

We should like to thank our colleagues at the Royal Military College of Science, Shrivenham, the School of Signals, Blandford and the other communications establishments and firms who have supplied material for this book.

Shrivenham. March 1983

Tony Wilcox
Mike Slade
Peter Ramsdale

Contents

List of Illustrations

1.

Command, Control and Communications (C^3)

INTRODUCTION

In any battle theatre, a commander must be able to command and control his forces so that his fighting assets are optimised. Historically, a clear dividing line was drawn between the command, control and communications systems, and all kinds of weapon systems. In a modern battle, movement of weapon systems is fast which means that reactions must be equally fast. Such speed can only be obtained if commands are passed swiftly to the elements under the commanders control. This depends on good radio communications on which commanders are becoming increasingly reliant.

Automatic Data Processing (ADP) facilities are increasingly being provided to assist military staff in the exercise of their command, control and intelligence (C^2I) duties. Communications provide reliable access to these ADP facilities and maintain the accuracy and mutual consistency of the distributed data bases on which operational decisions are based.

This book concentrates on communication systems which must provide the backbone of a successful coordinated command, control and communications (C^3) policy.

ELEMENTS OF C^3

Cybernetics is the theory of control and communications processes in animals and machines. Command and Control in the military context is concerned with the control of events and processes through the transmission and receipt of messages, and the concepts of cybernetics are directly relevant. A commander on the battlefield exercises command and control through a number of cybernetic or feedback loops which contain the following elements:

a. Surveillance. The commander will wish to have available to him intelligence from a large number of individuals and from a whole range of sensors.

b. Communications. Data for the commander and his staff can only reach them if suitable communication systems are provided.

c. Data Processing and Management. The raw data or information coming into a headquarters must be passed to those who need it, filtered, processed and then displayed in some suitable format to the commander or staff officers who must take action on it. Automation of the processes is an increasing trend in field force headquarters.

d. Decision Making. This is a matter for the commander aided by his staff. The organisation of the headquarters, its internal communications, and the aids provided for decision making are all important elements of this activity.

e. Communications. This is the only element to occur twice in the cybernetic loop, and a communication system to convey orders that result from the decision making process is just as important as communications to bring information into the headquarters. The communications systems can thus be fairly described as doubly important to command and control as they must convey commands as well as return the control information.

f. Action. The purpose of surveillance, ADP, communications and all the other constituents of the cybernetic loop is to indicate action, the end product of a command and control system.

It is clear from the analysis of command and control that communications are the vital element in the loop. The user must therefore understand enough about his communications to enable him to get the best out of them. The communications systems themselves must be survivable, flexible, reliable, secure and interoperable if they are to contribute positively to effectiveness in battle.

SYSTEMS APPROACH

As advances are made in technology, individual items of equipment become more complex. It becomes increasingly important that those involved as users of telecommunications, or as part of an equipment procurement organisation, are able to view a system as a whole. They must be able to assess critically the system capabilities without becoming involved in design detail. This book has this approach, and systems are studied rather than individual equipment details. However, to be able to assess a system, some technical knowledge of the principles involved is necessary.

The function of a communication system is to pass information accurately from one place to another. In the past the terminal points of a communication system were human beings but it is now not uncommon, with the advent of ADP and automatic weapon control systems, for the terminals to be machines. The measure of the effectiveness of a communication system is its efficiency in passing information and this is described by the following parameters:

a. Accuracy. This is measured in terms of speech quality or error rate in data systems.

b. Quantity of information passed.

c. Speed of passing information.

These parameters are measurable and are specified as the basic design criteria of a system, and the final form of the system depends on the emphasis placed on each.

COMMUNICATION SYSTEM

A generalised communication system, as shown in Fig. 1.1, has the following components:

a. Information Source. This produces a message which may be written or spoken words, or some form of data.

b. Transmitter. The transmitter converts the message into a signal, the form of which is suitable for transmission over the communication channel.

c. Communication Channel. The communication channel is the medium used to transmit the signal from the transmitter to the receiver. The channel may be a radio link or a direct wire connection.

d. Receiver. The receiver can be thought of as the inverse of the transmitter. It changes the received signal back into a message and passes the message on to its destination which may be a loudspeaker, teleprinter or computer data bank.

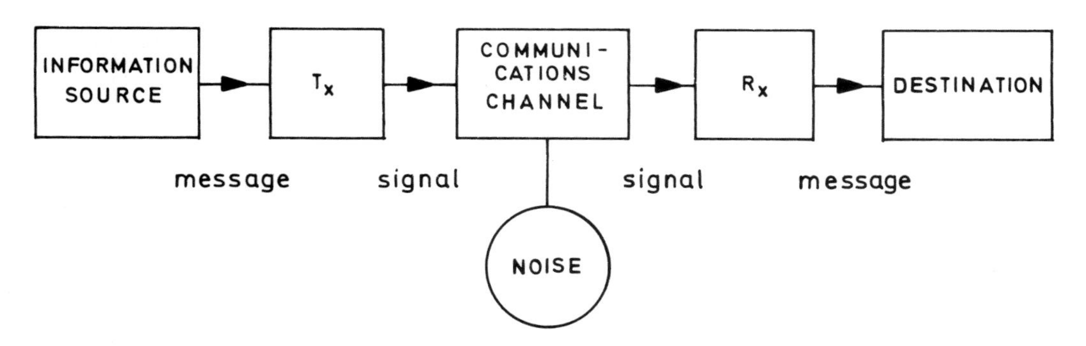

Fig. 1.1 Communications system

An unfortunate characteristic of all communication channels is that noise is added to the signal. This unwanted noise may cause distortions of sound in a telephone, or errors in a telegraph message or data.

ELECTROMAGNETIC SPECTRUM

The military communications electromagnetic spectrum is reproduced in Fig. 1.2.

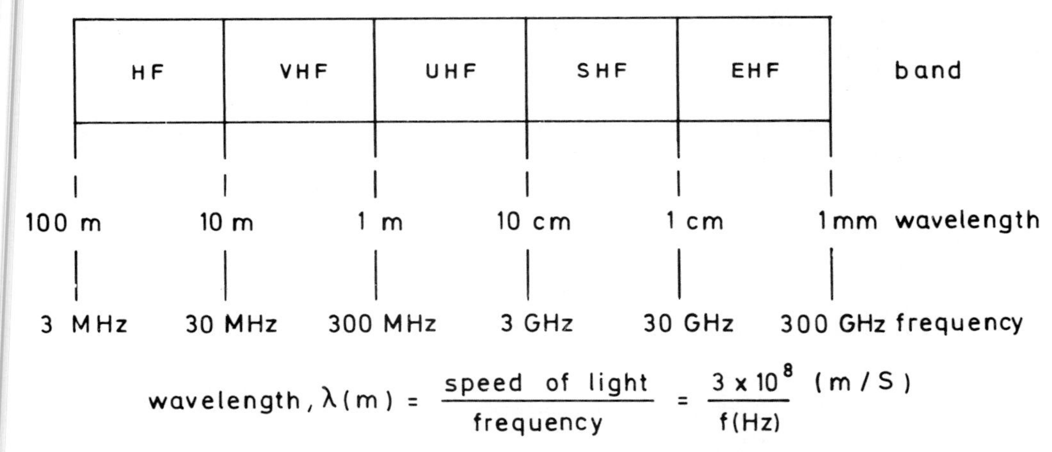

Fig. 1.2 Communications electromagnetic spectrum

It must be considered the key resource in telecommunications. The parts of the spectrum available for communication have been extended to shorter and shorter wavelengths partly because the spectrum is becoming very crowded with users at its longer wavelengths. Military users of the electromagnetic spectrum are restricted to certain frequency bands to fit in with the demands of civilian users. Even within the military bands there is intense competition between C^3 systems and other users such as guided weapons systems, surveillance devices, ships and aircraft for use of the limited number of frequencies available. The enemy also uses the same bands of the spectrum for his systems but additionally may use the spectrum destructively to hinder our use of it. Thus frequency allocation is an increasingly difficult exercise. Control of power outputs, new communication techniques, better designed equipment and the use of higher frequencies, up to the optical bands are measures taken to cope with demand.

OUTLINE

The fundamental concepts of radio communications are introduced in Chapter 2 and the basic elements of a simple communications system are described. These concepts are used and extended in Chapter 3 which is devoted to net radio. Both tactical and technical problems are addressed. Trunk radio systems are explained in Chapter 4. Various command structures and communication network principles are considered. The all important topic of electronic warfare is covered in Chapter 5 in which both tactical and technical approaches are discussed. There are many new communications techniques being investigated that will undoubtedly have an impact on future military communications. These have been gathered together in the final chapter in which their implications on both strategic and tactical communications are considered.

Good C^3 systems are an essential feature of a well organised modern army. Flexibility and rapid speed of response are achieved by the communications systems deployed. Although C^3 is a potent force multiplier, particularly when linked to weapons systems it does have weaknesses which when exposed can be exploited by a determined and knowledgeable enemy. As such it must be protected by tactical and technical means.

SELF TEST QUESTIONS

QUESTION 1 What problems do communicators face when a level of command is removed?

Answer ....

....

....

QUESTION 2 List the sensors that need communications support to optimise the C^2I duties of the staff.

Answer ....

....

QUESTION 3 Why is military communications frequency allocation becoming an increasingly difficult exercise?

Answer ....

....

ANSWERS ON PAGE 133

2.
Basic Radio

INTRODUCTION

System Concept

The term 'system' is very commonly used today to describe a variety of scientific, engineering or sociological situations. Since this particular book is biased strongly towards a system approach it is appropriate to begin by explaining this.

Railways, airlines, power grid networks, telephone networks, etc are a few examples of modern services and each one must be regarded as a system. If each service is broken down into its component parts then we are left with the basic electrical, mechanical and aerodynamical components which have been evolved over the years. In the past an improved device usually opened up new or improved applications and staff were trained by studying the device in depth at a fundamental level with the aim of understanding or improving its use in service. Today most services are so vast and inter-related that the introduction of a faster train, or a larger aeroplane, or a lighter power cable or a smaller telephone relay might in fact worsen the service as a whole because other factors have been overlooked.

The need to study the service as a whole is therefore very apparent and this is what is implied by system design or system study. Looked at in this way we need only consider components or devices in as much as they affect the overall aim of the system. This does not mean that fundamental studies are now irrelevant: on the contrary, one chooses the approach best suited to ones requirements. A component designer will probably favour the fundamental approach while a designer of overall services will be biased towards systems, the two approaches can never be divorced.

The general approach to the design of a communication system is in principle no different to that for a system of any kind. Various desirable and undesirable aspects of the behaviour of the system may be quantified and the system may be optimised using a computer and employing the methods of operational analysis. It must be stressed, however, that the system approach does not represent an easy way out. The topic often requires a high mathematical and scientific expertise and it is now regarded as a subject in its own right.

Communication Systems

A communication system can be described in detail by circuit diagrams and performance documentation. From such information it is often difficult to obtain an overall view of the function of the equipment. A system diagram is an attempt to overcome this difficulty in that it is a diagrammatic summary of the circuit functions. The notation often used is shown in Fig. 2.1.

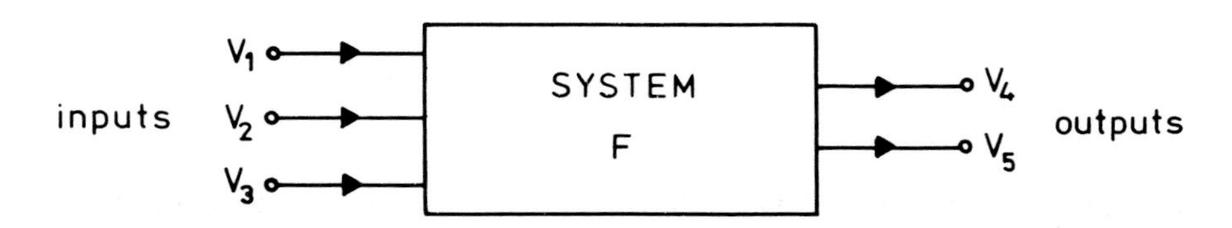

Fig. 2.1 System diagram

The system function F, acts on the inputs (excitations) V_1, V_2, V_3 to cause the outputs (responses) V_4, V_5. Arrows indicate the direction of signal flow. An example of a system often used in telecommunications is the amplifier shown in Fig. 2.2.

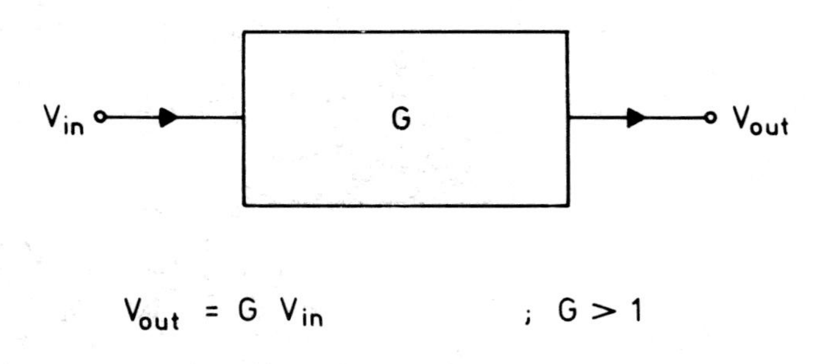

$$V_{out} = G\ V_{in} \qquad ; \quad G > 1$$

Fig. 2.2 Amplifier

The amplifier could be of any type using valve or transistor technology at either audio or radio frequencies. Typically the gain G could vary from 1 to several hundreds.

It is often more convenient to describe the operation of a system in terms of frequency rather than time, these being referred to as the frequency and time domains, respectively. Both descriptions convey the same information, as shown in Fig. 2.3.

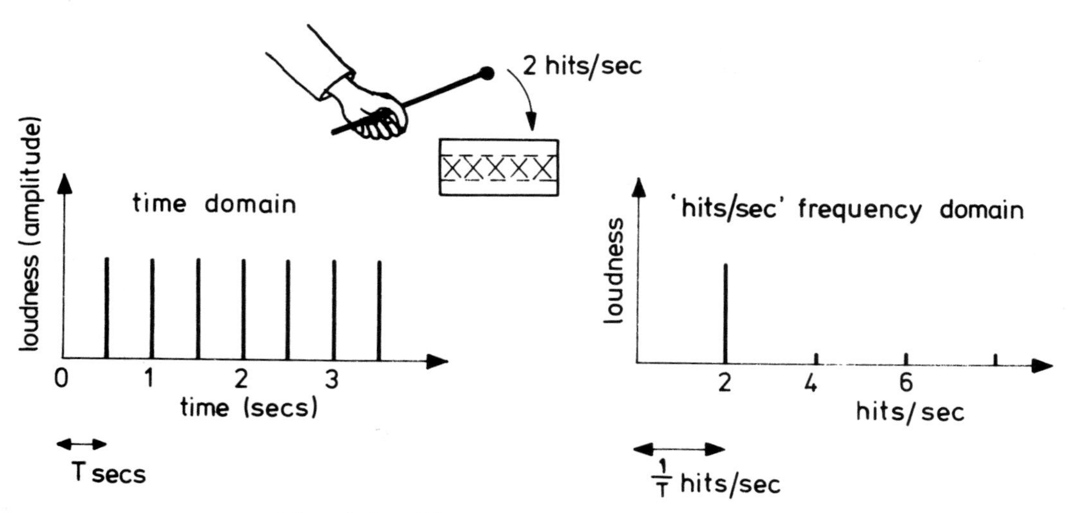

Fig. 2.3 Time and frequency domains

The time-waveform which is the building-block of telecommunications is the sine wave. Its time and frequency domain representations are shown in Fig. 2.4.

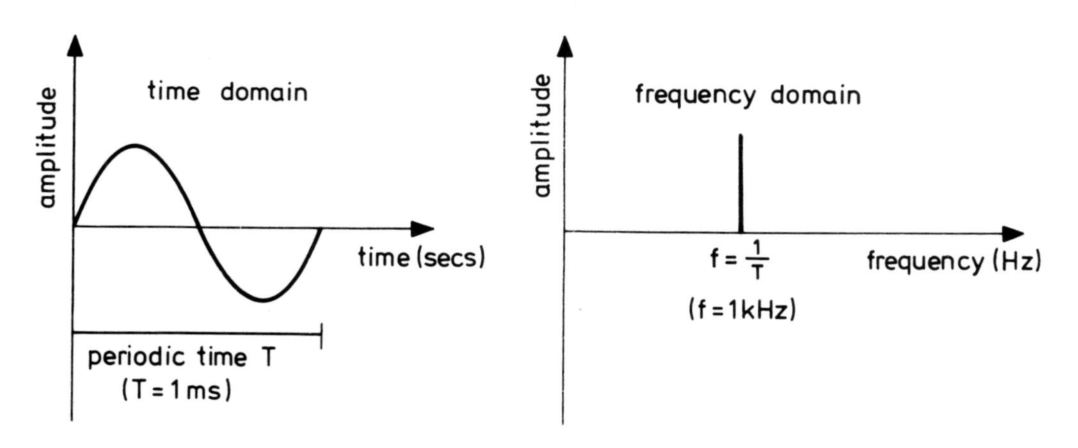

Fig. 2.4 Sine wave representations

The two most important characteristics of a sinusoidal waveform are its amplitude and its frequency, measured in Hertz (cycles/sec).

More complicated signals are often formed within communication systems. A simple example is the addition of two equal amplitude sine waves of different frequency. The resulting time and frequency domain representations are shown in Figs. 2.5 and 2.6 respectively.

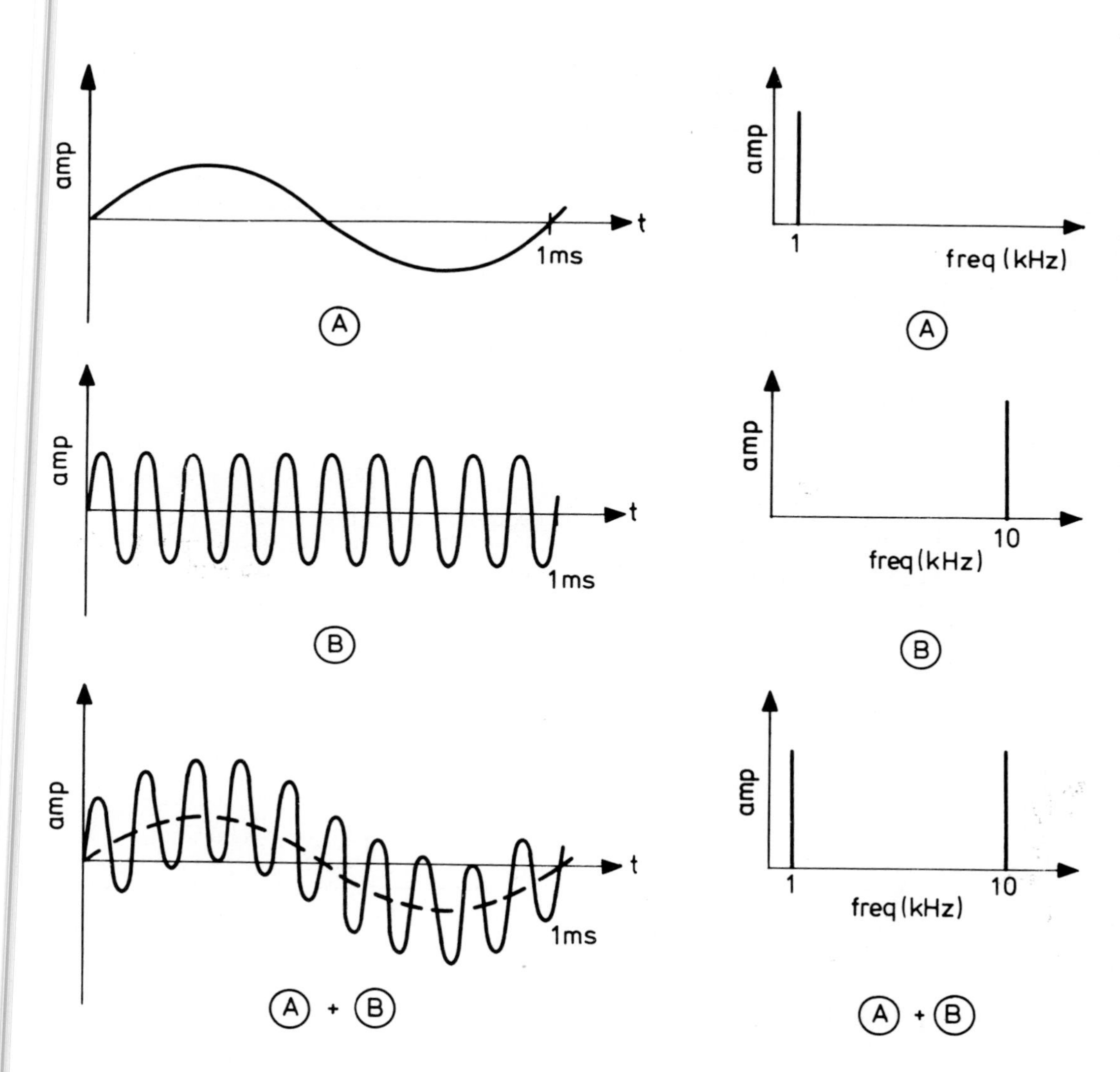

Fig. 2.5 Time domain representation

Fig. 2.6 Frequency domain representation

SIGNALS

All applications of electronics are concerned with the rapid handling of information; electrical signals are used to gather information as in radar, to send information as in telecommunications, or to sort information as in computers. Thus an appreciation of how electrical signals can represent information is essential. Once the nature of the signals is understood, whether they exist as radio waves or as the flow of electric current in wires, the ways of generating and processing these signals and transmitting them from point to point can be considered.

Waveforms

An electrical signal is a voltage or current which varies with time, often extremely rapidly. If this is expressed as a graph, a characteristic shape or waveform is revealed. It is vital to be able to study these waveforms, and this can be done using the cathode ray oscilloscope (CRO) which may essentially be regarded as a graph plotting device, normally used to show how voltage varies with time.

The heart of the cathode ray oscilloscope is the cathode ray tube illustrated in Fig. 2.7. Electrons are emitted by a hot cathode, focused into a fine beam and

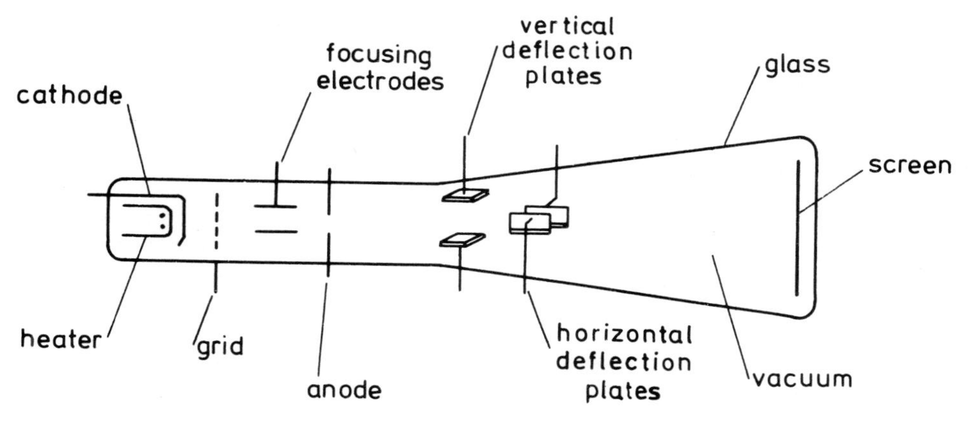

Fig. 2.7 Cathode ray tube

accelerated towards a screen by a high voltage. The point of impact becomes visible due to the fluorescence of the screen. The beam can be deflected to any part of the screen by applying a voltage to plates between which the beam passes.

A time base is produced by applying to the horizontal deflection plates, a voltage which increases linearly with time. The voltage variation to be observed is

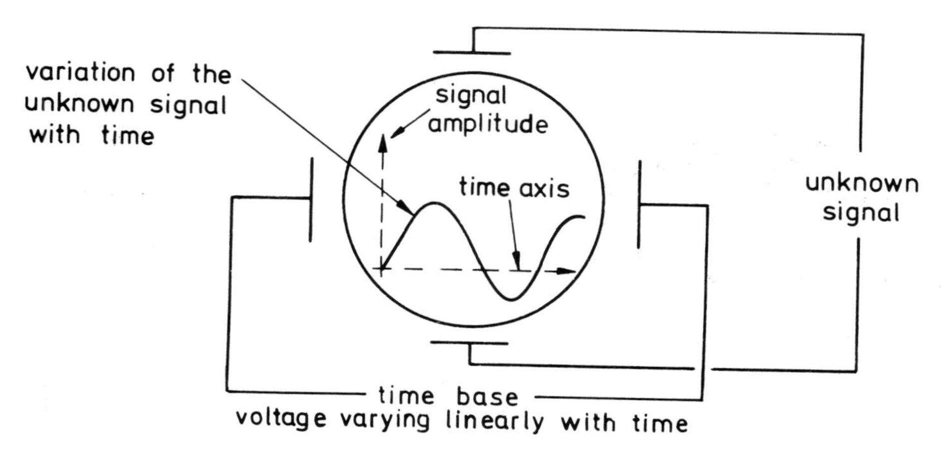

Fig. 2.8 CRO display

applied to the vertical deflection plates and the beam produces a visible waveform on the screen such as is shown in Fig. 2.8. The waveform simply shows how the signal varies with time.

The more complex voltage from a microphone can be examined using the CRO and a typical display of high and low frequency components can be seen in Fig. 2.9.

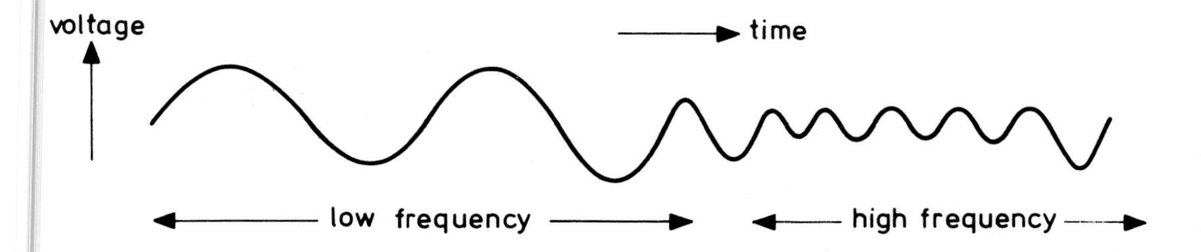

Fig. 2.9 Waveform with high and low frequency components

The highest frequency components in speech extend well beyond 3000 Hz. However, these are not vital to speech intelligibility and are not usually transmitted over a telephone circuit. For military quality voice communication only frequencies up to 3000 Hz are used.

There are numerous other types of waveform, for example, in radar and some telecommunication systems, pulse waveforms are used, as shown in Fig. 2.10. A useful feature of the CRO is that it can act as a clock which accurately measures the very short periods of time which may separate such pulses. It is easy to measure periods of time which are shorter than a microsecond (10^{-6} second) by this means.

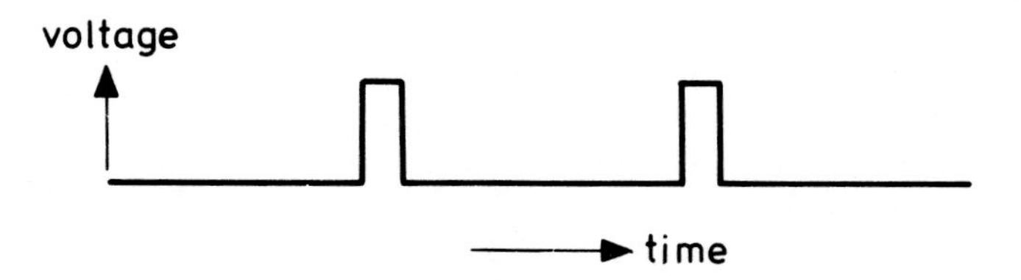

Fig. 2.10 Pulse waveform

An important theorem states that any complex waveform can be expressed as the sum of a number of simultaneous sinewaves of different frequencies. This enables us to describe any waveform by an amplitude-frequency graph known as the spectrum of the signal. For instance the complex wave of Fig. 2.11 may be expressed as the sum of 4 kHz, 5 kHz and 6 kHz sine waves and this can be checked easily by a graphical construction.

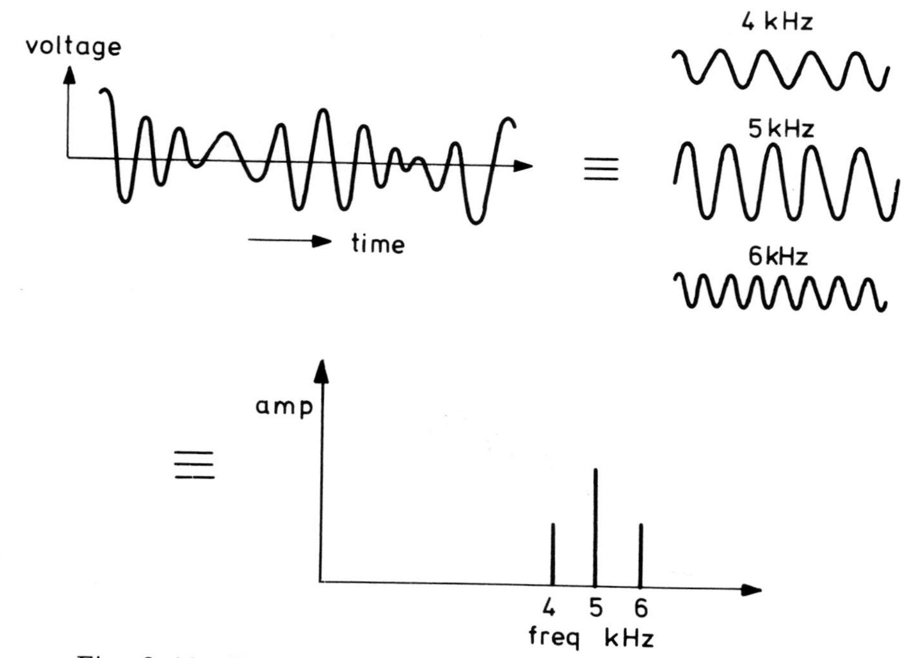

Fig. 2.11 Frequency components of a complex waveform

The waveform of a radio signal is always complex, but the array of frequencies in the spectrum will be contained within a finite band, the bandwidth of the signal. A radio receiver tuning dial is set to a frequency at or near the centre of this band. It receives the radio signal without distortion provided that it is equally sensitive to all frequencies within the signal bandwidth.

Bandwidth

If a tuning fork is struck then we are accustomed to hearing a pure audio tone which probably corresponds to a note on a musical scale giving a single line in the frequency spectrum as shown in Fig. 2.12.

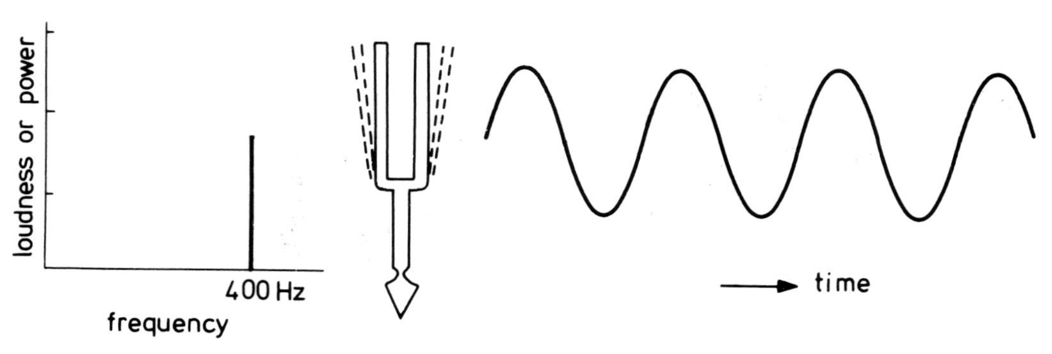

Fig. 2.12 Tuning fork oscillation

The ear, however, is not very sensitive to what takes place when the tuning fork is struck and delicate measurements will show that if the fork is struck at time t = 0, a wide spectrum of other frequencies occurs momentarily around the 400 Hz frequency. By the time t = 0.5 seconds only the 400 Hz tone is likely to be detected. If the fork is abruptly turned off then a transient spectrum will again appear; a good rule is that the slower the build up or decay of oscillations the less dominant are these transients. An every day demonstration of this is a light switch which in switching on the a.c. mains waveform may create a transient spectrum which interferes with radio and TV over a wide range of frequencies.

We can now be curious and ask what happens if the fork is turned on and off so abruptly and frequently that these transients do not have time to die away. The resultant frequency and time plots are shown in Fig. 2.13. On listening to the fork we will find that the tone has become very harsh due to the additional frequency components.

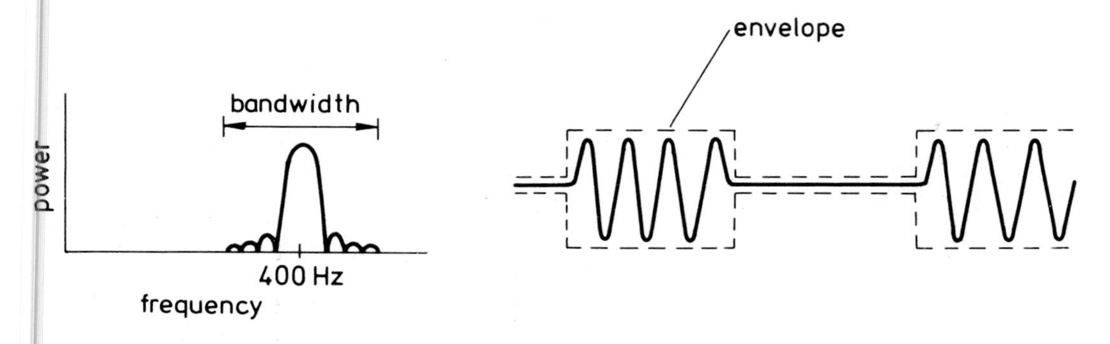

Fig. 2.13 Modulated tuning fork oscillation

Noise

There is much talk today about the high acoustic noise level from cars, lorries and aircraft. A parallel situation exists for electrical signals. A radio signal is generally very weak when it arrives at the receiver and the electrical background noise can ruin reception. Natural noise has both a component generated within the atmosphere around the earth and a galactic component from space. Man made, electrical equipment noise includes interference from car ignition, light and power switches, motors and poor connections within electronic apparatus. An example which is of particular military importance is the electrical noise from sparks created by the movement and vibration of a vehicle body and chassis. This makes a radio more difficult to use when moving. Also because any body generates noise in proportion to its physical temperature, there are noise contributions from all the components of a receiving system and thermal noise from the ground. Figure 2.14 is a sketch of the frequency bands in which these effects appear.

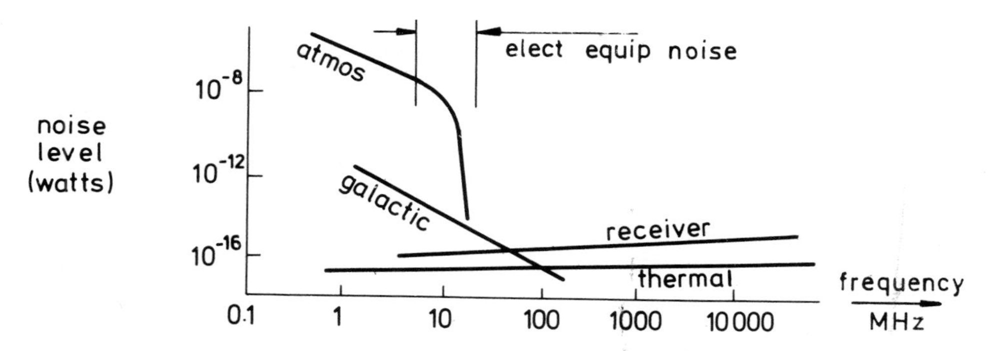

Fig. 2.14 Frequency bands for various noise effects

Decibels

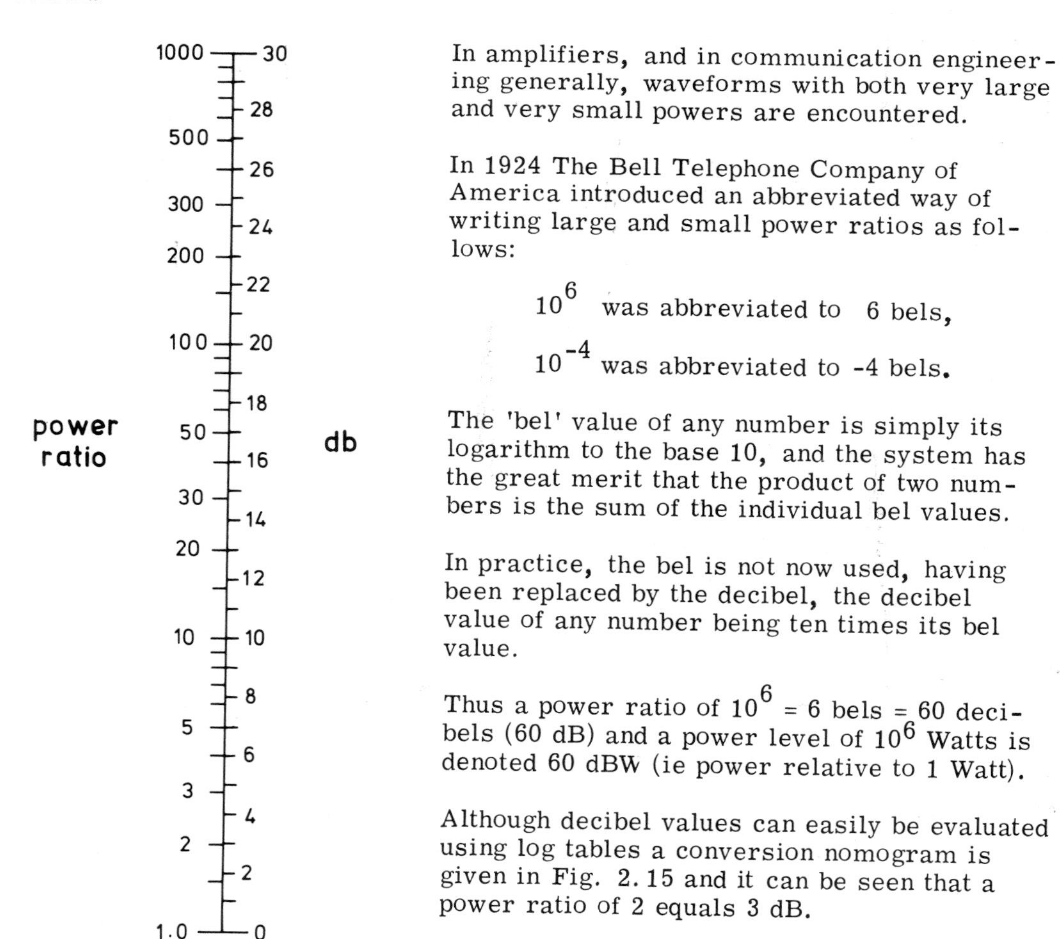

In amplifiers, and in communication engineering generally, waveforms with both very large and very small powers are encountered.

In 1924 The Bell Telephone Company of America introduced an abbreviated way of writing large and small power ratios as follows:

10^6 was abbreviated to 6 bels,

10^{-4} was abbreviated to -4 bels.

The 'bel' value of any number is simply its logarithm to the base 10, and the system has the great merit that the product of two numbers is the sum of the individual bel values.

In practice, the bel is not now used, having been replaced by the decibel, the decibel value of any number being ten times its bel value.

Thus a power ratio of 10^6 = 6 bels = 60 decibels (60 dB) and a power level of 10^6 Watts is denoted 60 dBW (ie power relative to 1 Watt).

Although decibel values can easily be evaluated using log tables a conversion nomogram is given in Fig. 2.15 and it can be seen that a power ratio of 2 equals 3 dB.

Fig. 2.15 Nomogram relating the power ratio to the decibel

MODULATION

Modulation is the method by which the information waveform is imposed onto a radio frequency carrier. The topic of modulation is often difficult to grasp because there are many alternative methods and it is often not obvious why one particular method is used rather than another. The approach used here is to describe first the most common modulation method which is Amplitude Modulation (AM) and show how it is applied in an elementary radio receiver. This is followed by a description of Frequency Modulation (FM) which is the more widely used method at VHF and above.

A British Army CLANSMAN VRC-321 radio radiates a frequency in the range 1.5-30 MHz in order to carry speech, although telephone quality speech waveforms are confined to frequencies below 3 kHz. In a simple telephone line circuit the speech waveform from the microphone is sent directly to the destination in its original form. In radio however, we cannot do this for two important reasons:

a. Speech cannot be launched directly as a 3 kHz radiated wave from an antenna of a practical size because for reasonable efficiency, whip antennas are at least a quarter wavelength. In this case it would be 25 km long!

b. Even if such a wave could be launched, the operational usefulness would be severely limited by a complete inability to separate one conversation from another. It would be just as if all telephones were connected to the same pair of wires.

Amplitude Modulation

An unchanging sine wave does not convey any information and so in AM the amplitude of the carrier sinewave frequency f_c is varied in sympathy with the amplitude of the information signal of frequency f_s as shown in Fig. 2.16.

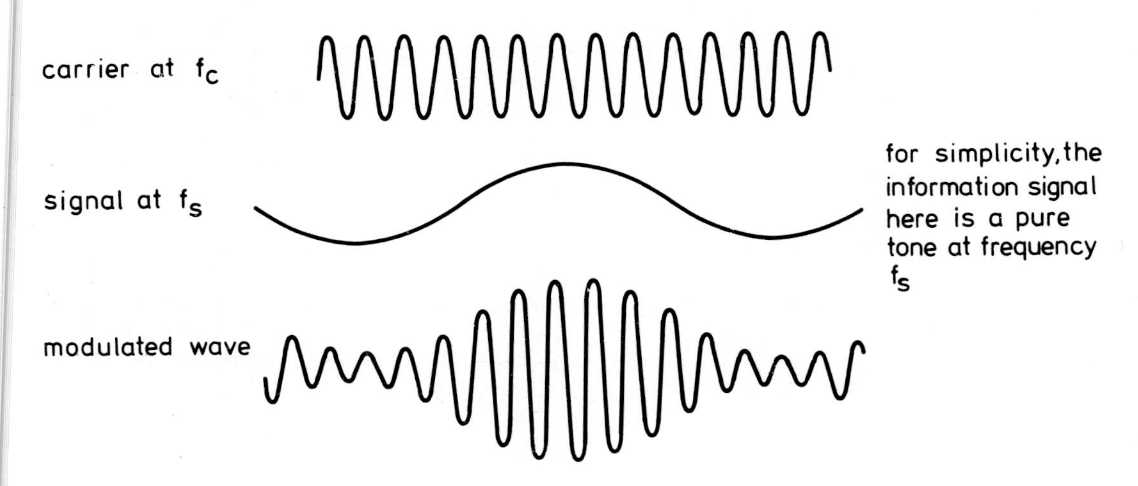

Fig. 2.16 Amplitude modulation

In the case illustrated, the complex modulated wave is, in fact, the sum of three sinewaves at frequencies f_c, $(f_c + f_s)$ and $(f_c - f_s)$. These are called the carrier, upper-side frequency and lower-side frequency and this spectrum is shown in Fig. 2.17.

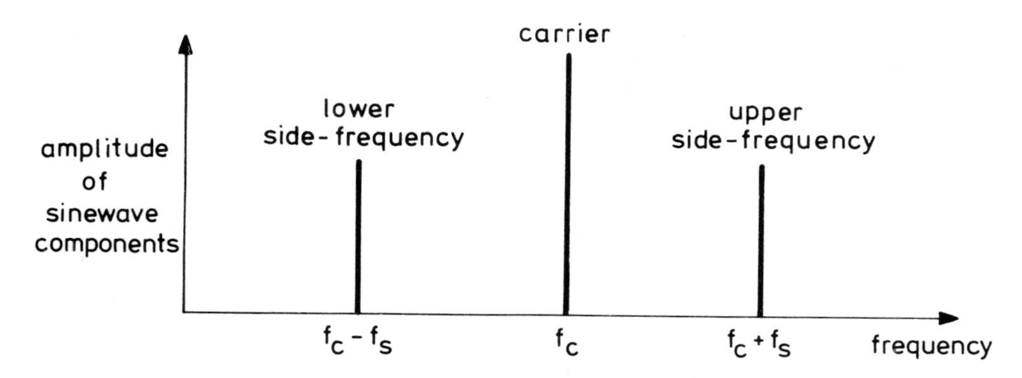

Fig. 2.17 Spectrum of amplitude modulated wave

Now consider the case of a practical modulating signal such as speech which contains a band of frequencies so that f_s varies between 0 and 3 kHz. The spectrum of the corresponding modulated carrier wave now occupies a bandwidth of 6 kHz as shown in Fig. 2.18. To receive this signal, the receiver must accept all the frequency components in a band 3 kHz above and below the carrier.

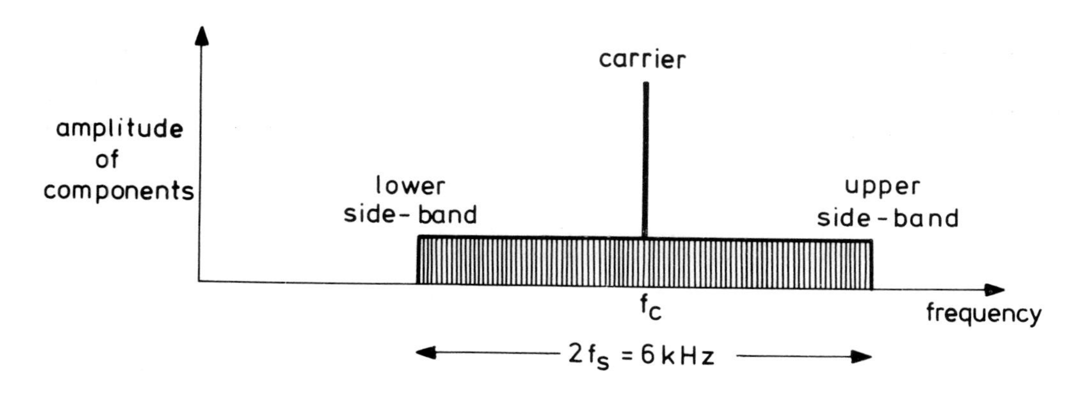

Fig. 2.18 Spectrum of practical AM signal

An elementary radio receiver can be made as shown in Fig. 2.19. Although several transmissions are received by the antenna only a 2 x f_s bandwidth is amplified within the receiver. The demodulator then extracts the information from its associated carrier.

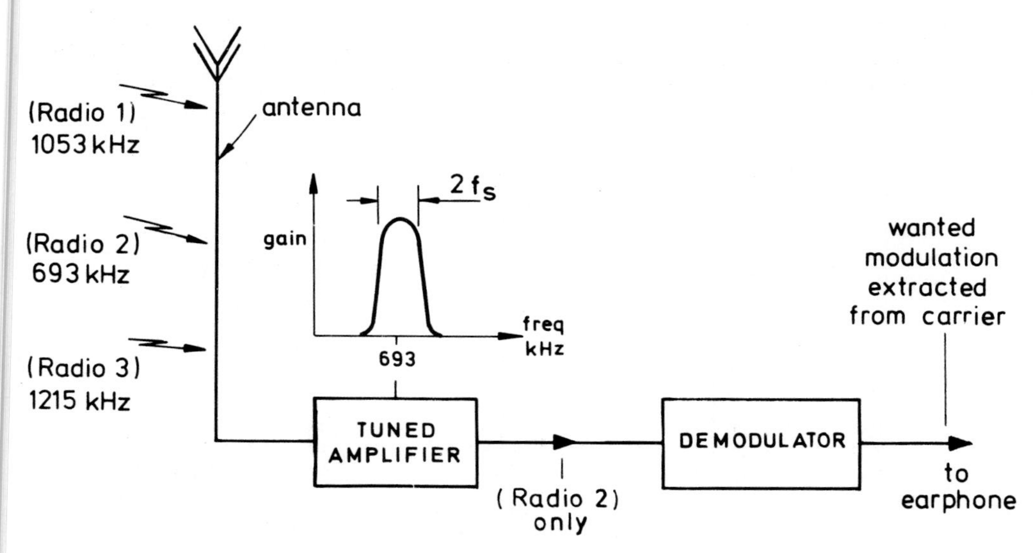

Fig. 2.19 Elementary radio receiver

In normal AM it is found that at least two thirds of the power is at the carrier frequency, it does not contain any information and it is not affected by the modulating signal. Hence the carrier is sometimes suppressed so that the transmitter power is used more effectively. Unfortunately, this Double Side Band Suppressed Carrier transmission requires a much more complicated and expensive demodulator at the receiver. In addition, both sidebands of AM contain the same information and so it is really only necessary to transmit one of them. This variant is called Single-Side Band (SSB) and halves the required bandwidth but again leads to a more complex receiver. In most military applications saving bandwidth is more important than this added complexity and SSB is the preferred solution for current HF net radio (Chapter 3).

Frequency Modulation

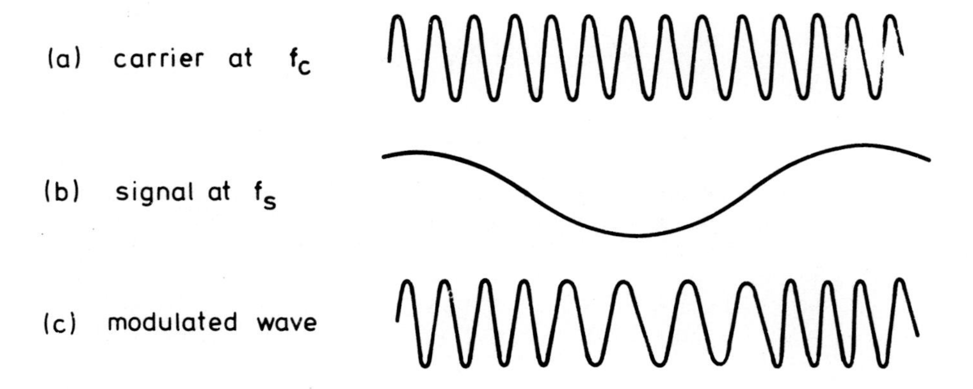

Fig. 2.20 Frequency modulation

In FM the frequency of the carrier is varied in sympathy with the amplitude of the information signal, as shown in Fig. 2.20. As the amplitude of the signal goes more positive the FM signal increases in frequency and vice versa. The frequency spectrum of an FM signal is very complicated. The number of significant lines in its spectrum can be very large. However, the greater the bandwidth used, the better the system performs in the presence of noise.

Figure 2.21 shows how the carrier frequency varies in response to the changing information signal amplitude. The maximum excursion of the carrier is known as the frequency deviation, Δf.

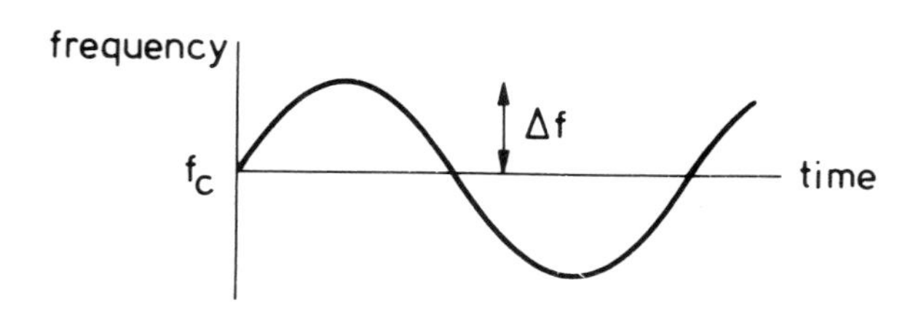

Fig. 2.21 Frequency deviation

The important properties of FM can be described in terms of the modulation index, m and this is found from:

$$m = \frac{\text{frequency deviation}}{\text{maximum information signal frequency}} = \frac{\Delta f}{f_s}$$

and as Δf can be much greater than f_s, m may be greater than 1. Compared with SSB the improvement in the ratio of wanted signal to interfering noise is $\frac{3}{2} m^2$. It is achieved at the expense of extra bandwidth which becomes approximately $2(m + 1)$ times the bandwidth of the original information signal. The UK VHF-FM broadcast signal, for example, occupies 180 kHz (f_s = 15 kHz, m = 12) but the high quality sound it can reproduce is so much better than AM that this apparently extravagent use of bandwidth can be justified. In CLANSMAN VHF net radios as described in Chapter 3, m = 1.67. In general, military sets use a fairly low value for m to economise in bandwidth. There are other modulation techniques and it is found that those which use wider bandwidths give better protection from noise and interference.

INFORMATION THEORY

Information theory deals with any flow of information. Its application to telecommunication systems provides an explanation of the relationships between signals, noise and bandwidth. Although the theory is somewhat idealised it has put telecommunications on a firmer footing and, indirectly, inspired the design of better equipment.

A fundamental result is that

$$\text{Bandwidth} \times \text{Time} = \text{Constant}$$

Thus the bandwidth required can be halved by taking twice as long to send a message. Another technique for saving bandwidth is to reorganise the message by a coding method to eliminate redundant content. For example, a single code word sent by teleprinter requires less bandwidth and time than its equivalent clear text spoken message. Although redundancy of information is seen as a waste of channel capacity it does provide some safeguard against errors due to noise interference.

The capacity of a channel is the speed, in binary digits/second, at which information can be sent. Shannon's formula relates the error free capacity C of a channel to its bandwidth W, interfering noise power N and signal power P.

$$C = W \log_2 (1 + P/N) \text{ bits/sec}$$

Shannon's formula is a useful guide in system design. For example, if the radio bandwidth W and noise level N are fixed, any capacity C can be achieved by increasing the signal power P. Alternatively C may be held constant in higher noise (N) conditions by increasing W. This is the principle behind broadband modulation systems such as FM and various techniques for overcoming jamming (see Chapter 5).

ANTENNAS

The design of communications antennas is a compromise between electrical efficiency, physical size, cost and simplicity. Some, particularly mobile antennas, are inefficient as radiators but represent the best compromise solution. Any improvements to efficiency that can be made to antenna systems are well worthwhile, since these represent an effective increase in transmitter power or receiver sensitivity at no cost in terms of the radio equipment. For ease of matching to the radio and for high efficiency, whip antennas should ideally be $\frac{1}{4}$ wavelength.

Antennas can be used for both radiating and receiving energy. The polar diagram of an antenna shows the relative powers radiated in different directions. For a vertical whip the polar diagrams are drawn in Fig. 2.22.

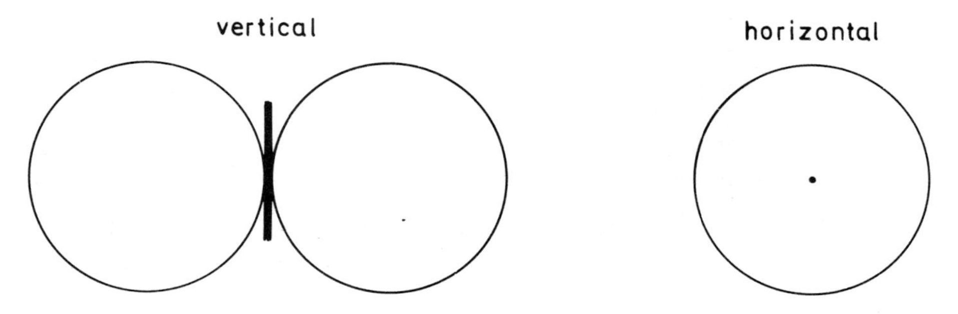

Fig. 2.22 Polar diagrams of whip antenna

The whip gives omnidirectional coverage in the horizontal plane. By using additional radiating elements the antenna can be made to transmit more power in any desired direction.

The gain G of an antenna is defined by

$$G = \frac{\text{power density in direction of maximum radiation}}{\text{power density at same distance from an isotropic antenna radiating the same total power}}$$

where an isotropic antenna is one that radiates equally well in all directions.

Omnidirectional Antennas

Omnidirectional antennas are used in net radio (Chapter 3) and include manpack and vehicle mounted antennas; they are usually end-fed whips called monopoles having a nominal length of $\frac{1}{4}$ wavelength. They can be considered as $\frac{1}{2}$ wave vertically mounted dipole antennas with the feed point at the centre. The lower half does not physically exist but is simulated, electrically, by the currents which flow in the material on which it is mounted (Fig. 2.23).

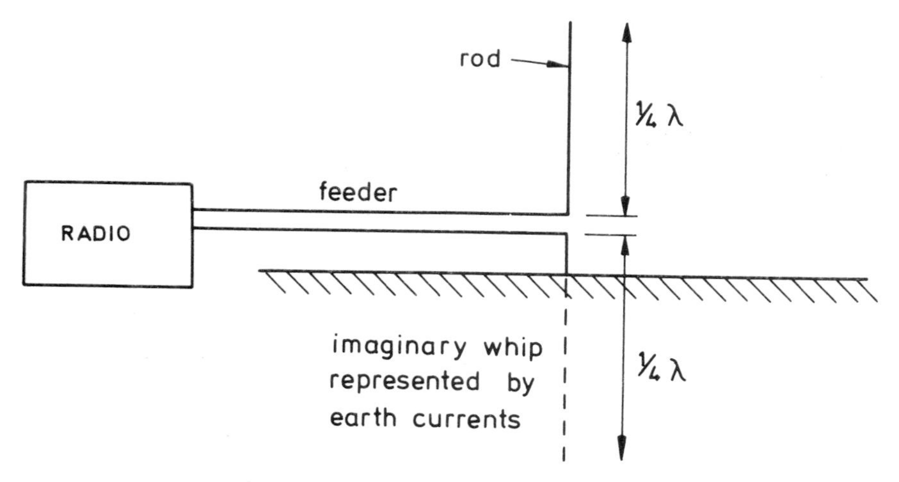

Fig. 2.23 End-fed whip antenna

For the antenna to work exactly as a dipole the ground plane must be a flat area of several wavelengths in extent and of perfectly conducting material. In practice this is never the case. On a manpack set the ground plane consists of the case of the set which is small and of poor conducting material. The ground plane of a vehicle antenna is the vehicle body which is better than in the manpack case but still far from ideal. The effect of these inferior ground planes is twofold. High losses in the material of the ground plane reduce the amount of energy available for radiation and the polar diagram of the antenna is distorted so that blind spots occur where there is reduced radiation in particular directions.

Since an antenna is only $\frac{1}{4}$ wavelength long at one particular frequency and radio equipments operate over a band of frequencies either the length of the rod must be varied for each frequency, which is not generally practical, or some method is required of making a fixed length antenna appear electrically as a $\frac{1}{4}$ wavelength whatever frequency is being used. This is carried out by some form of antenna tuning unit (ATU). In manpack radios this is usually built into the case and any controls required are mounted on it; for vehicle sets it is normally a separate unit (Fig. 2. 24) and may have to be adjusted at each frequency, although with some modern equipments, this is done automatically.

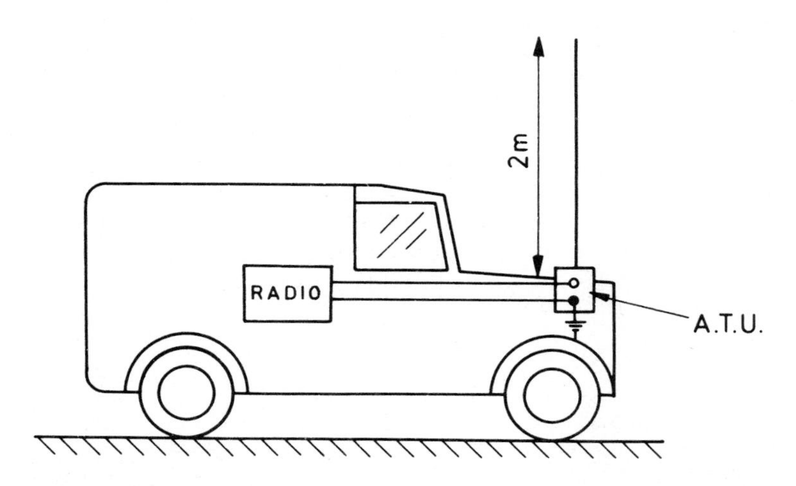

Fig. 2. 24 Vehicle mounted whip antenna with ATU

A manageable manpack rod is about 2. 5 metre and a vehicle rod is about 5 metre whereas at 5 MHz the $\frac{1}{4}$ wavelength is 15 metre. Since $\frac{1}{4}$ wavelength at 39 MHz is 1. 85 m, VHF whip antennas are generally more efficient than HF.

Directional Antennas

Directional antennas are used for UHF and SHF radio relay trunk communications (Chapter 4). The main advantages of directional antennas are first that by concentrating radiation in a particular direction, they give an effective power gain to the link. Secondly as receiving antennas, they give some protection against interference from unwanted stations on the same frequency; when transmitting they reduce interference to other stations.

As radio relay equipments use groups of frequencies, wideband antennas are required to prevent frequent retuning.

Typical antennas are the log-periodic (Fig. 2. 25), and the stacked dipole and corner reflectors shown in Figs. 2. 26 and 2. 27. These latter two antennas are used with the UK TRIFFID radio relay equipments covering the bands 610-960 MHz and 1350-1850 MHz respectively.

Fig. 2.25 Log periodic antenna

Fig. 2.26 Stacked dipole

Fig. 2.27 Corner reflector

An alternative dipole array antenna, the Yagi has more directivity but inadequate bandwidth for modern radio relay equipment.

Reflector Antennas

SHF frequencies are used for beyond-the-horizon communications via satellite and are now also being used in trunk radio. At these frequencies reflector antennas are the usual type. The reflecting surface focuses radiation to a small feed as shown in Fig. 2.28. The directional gain G, beamwidth θ_B and effective aperture A which is proportional to physical size, are related as follows:

$$G = \frac{4\pi A}{\lambda^2} \propto \frac{1}{\theta_B^2}$$

where λ is the wavelength. As the wavelength is small high gains and narrow beams are achieved.

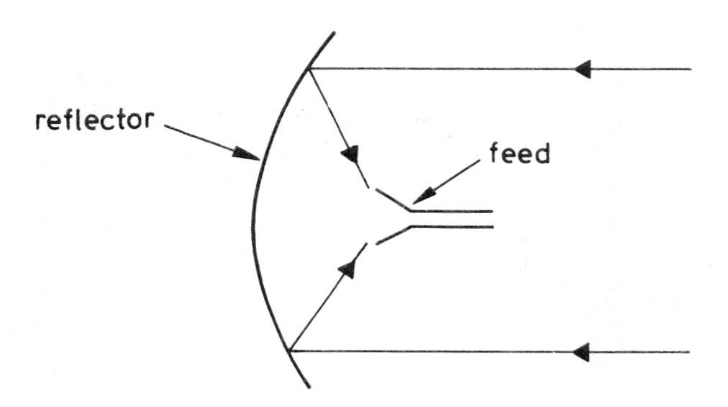

Fig. 2.28 Basic reflector antenna

PROPAGATION

Space Waves

Propagation of VHF and above is mainly by the space wave which consists of a direct wave and a ground reflected wave. The direct wave travels through the troposphere and is bent slightly downwards due to refraction by the atmosphere. The attenuation of such a wave is known as the free-space loss (FSL) and is expressed as

$$FSL = \left(\frac{4\pi r}{\lambda}\right)^2$$

where r is the distance and λ the wavelength. This takes into account the spreading of the radiation as it travels away from its source at the speed of light c (= 3×10^8 m/s). Frequency f is related to wavelength by the formula

$$c = f\lambda$$

If the antennas are close to the ground there will be a significant ground reflected wave. Whether a reflected wave reinforces or diminishes the direct wave depends on the difference in path lengths for the two paths. It is usually accepted, for practical link planning, that losses due to a reflected wave can be ignored if there are no likely reflecting surfaces within the first Fresnel zone, as shown in Fig. 2.29.

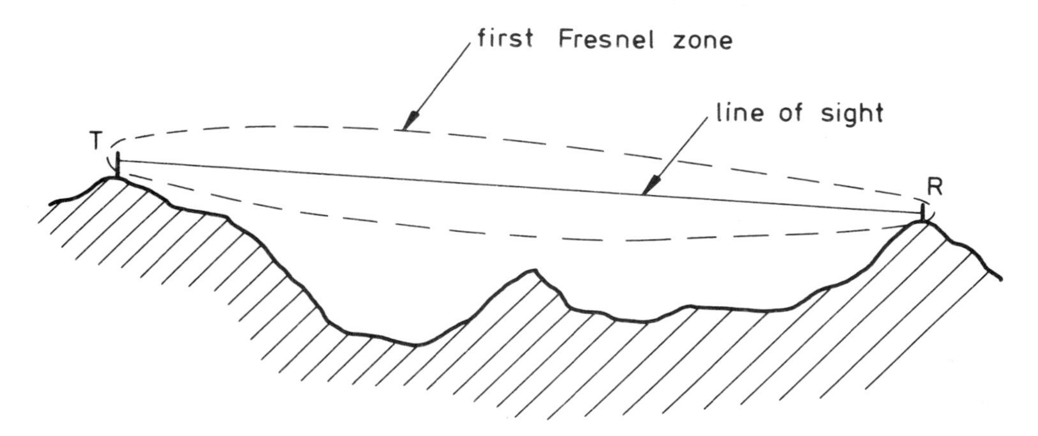

Fig. 2.29 First Fresnel zone clearance

Beyond-the-Horizon

Figure 2.30 shows possible mechanisms for radio propagation over-the-horizon.

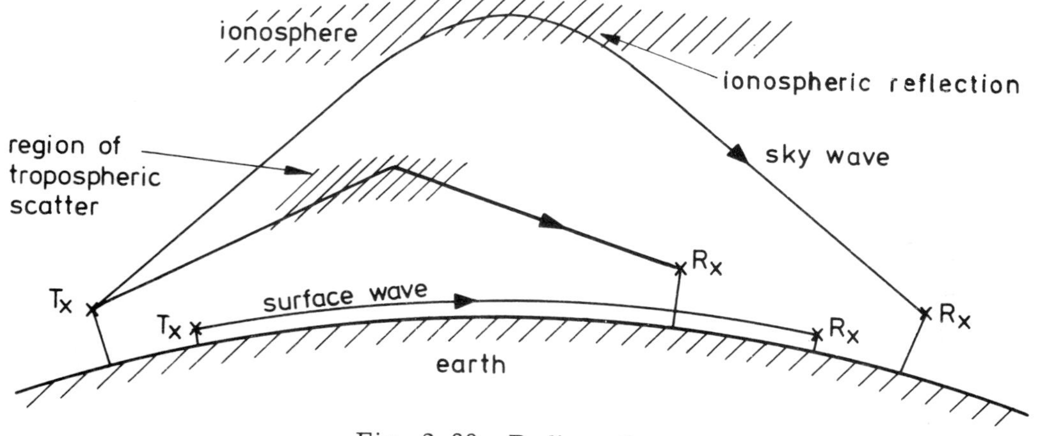

Fig. 2.30 Radio paths

Surface wave. Radio waves are able to follow the curvature of the Earth by travelling in the surface-wave mode. A surface wave induces electric currents in the surface over which it passes, with the result that the wave loses power, or in other words is attenuated. Above a few Megahertz this attenuation is very high and the surface wave becomes insignificant.

Ionospheric reflection. Between 50 km and 500 km above the surface of the Earth there are layers of air which are ionised by solar radiation. The intensity of ionisation and the heights of the layers vary with location, season and time of day.

The effect of ionisation is to bend waves reaching the layers. The bending produced increases with increasing intensity of ionisation or decreasing frequency. Hence at sufficiently low frequencies waves are bent and return to the surface of the Earth as if by reflection.

The higher layers are the most intensely ionised and maximum frequencies reflected by them vary from about 3 MHz to 30 MHz. The lower layers are ionised to a lesser degree and do not reflect but introduce attenuation.

The region of the reflection is continually moving and changing so that the reflected wave varies in intensity and phase. This effect is known as fading.

Figure 2.31 shows the general form of the ionised layers. For HF communications (3-30 MHz) the D and E layers attenuate the signals. The F-layer is more

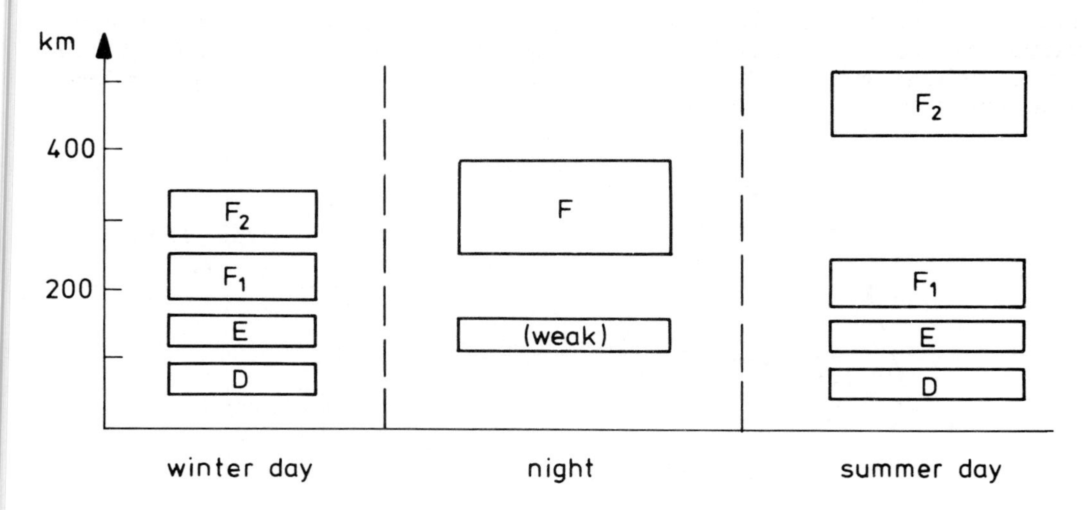

Fig. 2.31 Structure of ionosphere

intensely ionised and reflects the waves. Hence at night, when the lower layers are weak, many signals are reflected over great distances.

The highest frequency which is reflected at vertical incidence is known as the critical frequency f_c, and it depends on the intensity of ionisation of the layer.

For a given point-to-point radio link the launch angle θ, as shown in Fig. 2.32, is fixed by the height of the ionised layer and the maximum usable frequency f_{muf} is given by

$$f_{muf} = \frac{f_c}{\cos \theta}$$

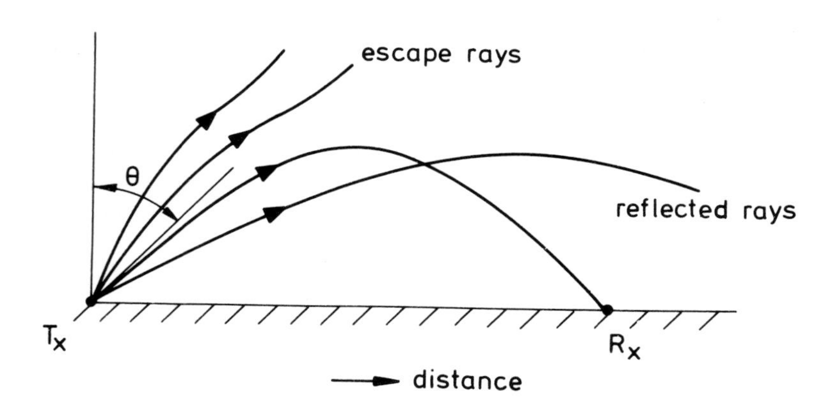

Fig. 2.32 Critical angle

In practice frequencies about 15% below f_{muf} are used to allow for ionospheric variations. Although much lower frequencies are possible, there is an additional fading problem due to multihop transmission, Fig. 2.33, and increasing attenuation through the D and E layers.

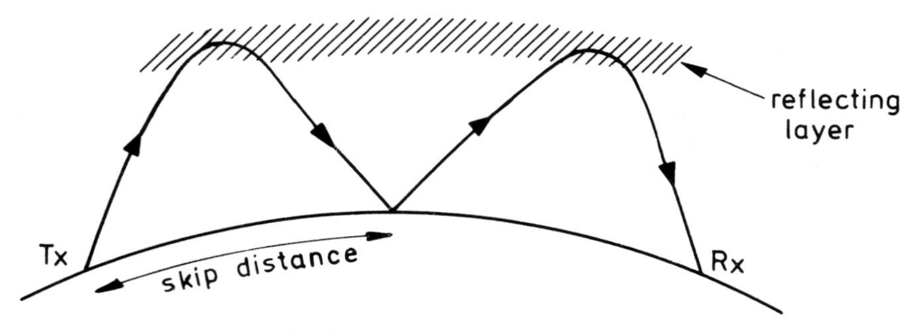

Fig. 2.33 Multihop transmission

For a given frequency there is a skip distance, which is the shortest range obtainable by ionospheric reflection at that frequency.

When a receiver receives two or more roughly equal waves which have travelled by paths of different length, fluctuations in the reflections can cause deep fades which may be frequency selective. The probability of a strong multiple-hop wave in a situation in which a single hop is required is reduced by choosing a frequency close to f_{muf}.

Tropospheric scatter. The lowest part of the atmosphere is called the troposphere and extends to a height of about 10 km.

Over-the-horizon propagation may be obtained by using highly directional antennas and directing both the transmitting and the receiving antennas towards the same region of the troposphere called the scatter volume, as shown in Fig. 2.34.

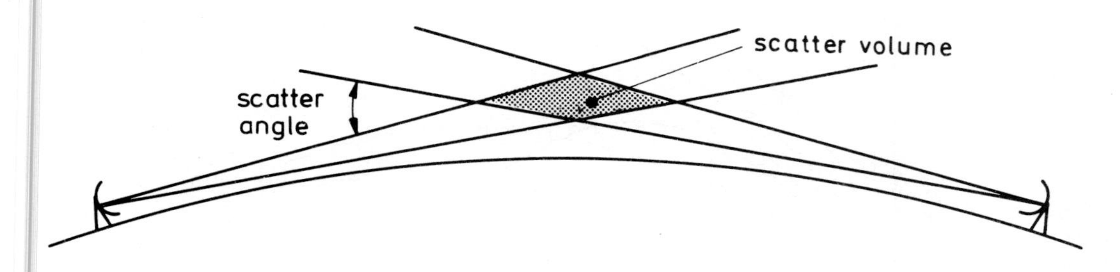

Fig. 2.34 Tropospheric scatter propagation

For scatter angles of the order of 1°, the path loss is typically 60-80 dB greater than it would be over a direct path of the same length. This loss increases rapidly with scatter angle. Hence it is important to minimise scatter angle by siting antennas to have a good view of the horizon.

Because of the very large path loss, high transmitter power is used, creating a serious radiation hazard close to the radiating antenna.

The path loss increases with frequency and therefore the choice of operating frequency is a compromise between the need to keep antennas small and the need to keep transmitted power low. It is also important to note that turbulence in the scatter volume gives rise to fading of the received signal. A typical use of tropospheric scatter systems is in high capacity, static, over the horizon links. Two examples of current military systems are shown in Table 2.1.

TABLE 2.1 Tropospheric Scatter System Parameters

	STARRNET	BERRNET
Frequency	2 GHz	4.5 GHz
Antenna diameter	18 m	9 m
Transmitted power	10 kW	1 kW
Service provided	60 voice channels	36 voice channels
Range	350 km	250 km

Urban Propagation

In an urban environment, the direct path between transmitter and receiver is, in general, obstructed by buildings. At UHF, waves which have travelled through buildings or have been diffracted around them are usually too weak to be of use and, therefore, communication is possible only be means of multiple reflections from the surfaces of the buildings.

The usual arrangement is to site a base station in a prominent position and to route all communications through that base station, not attempting to transmit directly from one mobile terminal to another.

Radiation from the base station illuminates the area by multiple reflections. A moving receiver in that area moves through a complicated interference pattern, experiencing deep fading as it passes through maxima and minima in the pattern. It is found that these variations are about half a wavelength apart. The frequency with which the fades occur depends upon the speed of the vehicle and the wave-length used.

The average received signal is typically about 50 dB below that which would be received over a direct path of the same length but varies with the general nature of the surroundings through which the vehicle passes.

Fading

Unpredictable fluctuations of the amplitude and phase of the received signal are an undesirable feature of all radio systems, but are particularly serious in over-the-horizon and urban systems.

Fading is often described by quoting the outage time and the fading rate. As its name suggests, outage time refers to periods in which the system is unusable due to the signal strength being too low for its reliable reception. The fading rate is the average frequency at which the received signal strength varies.

The fading rate depends on the propagation path. For example in over-the-horizon HF systems it is typically between 10 and 40 fades/minute. This is sufficiently close to syllabic rates to be a nuisance in telephony. In troposcatter systems it is typically 10 fades/sec. For a mobile receiver operating at 1 GHz and travelling at 50 km/hr the rate is about 100 fades/sec.

Complicated systems are used to reduce the effects of fading, they provide two or more signal paths, which are sufficiently different to be unlikely to fade simultaneously. Typical methods are to use two frequencies or several antennas. These methods are known respectively as frequency and space diversity, but because of their complexity, are normally found only in static systems.

SELF TEST QUESTIONS

QUESTION 1 Why are sine waves used to describe system response, when normally more complex waveforms are encountered?

Answer ....

....

QUESTION 2 What sources of electrical noise are likely to be encountered by a radio receiver on the battlefield?

Answer ....

....

QUESTION 3 Why is SSB amplitude modulation used?

Answer ....

....

QUESTION 4 What is the main advantage of FM over AM?

Answer ....

....

QUESTION 5 The signal-to-noise (P/N) in an ideal communications channel is 7, and the bandwidth W is 16 kHz, calculate the maximum information capacity C.

Answer ....

....

QUESTION 6 Explain why the performance of an antenna on a manpack is degraded.

Answer ....

....

QUESTION 7 What types of antennas are suitable for military trunk radio?

Answer ....

....

QUESTION 8 Calculate the loss in a 10 km line of sight propagation link at 50 MHz.

Answer ....

....

QUESTION 9 What forms of propagation can be used to give over-the-horizon communications?

Answer ....

....

QUESTION 10 What causes signal fading?

Answer ....

....

ANSWERS ON PAGE 133

3.
Net Radio

INTRODUCTION

Combat net radio is the primary means of command in the forward battle area. It has what is normally described as an 'all-informed capability'. This means that when one station transmits, the message is received by all other stations on the net, whether the message is of interest to them or not. An advantage is that general information about the state of the battle is available to all users of the net. Combat radio net uses a single frequency way of operating known as simplex, this means that transmissions can only take place in one direction at a time and only one transmitter can be on the net at any given time. This simplex mode creates a need for some form of radio procedure to ensure efficient use of the net so that only one user transmits at a time. This reduces the amount of time available to each user because a waiting period is needed to ensure that the net is free of other active transmitters.

These operational characteristics dictate the way net radio is used. It is of particular value at tactical levels and in mobile operations undertaken by formations and units below corps headquarters level. Single channel voice communication is the primary form of traffic and both HF and VHF bands are suitable.

Before discussing specific types of net radio, the operation of the standard forms of radio transmitter and receiver need to be outlined. The underlying principles of most radio stations are similar regardless of power or operating frequency.

TRANSMITTERS

The system diagram of a transmitter is shown in Fig. 3.1 and its component parts are as follows:

a. Master oscillator. This module is designed to produce a sinewave over the whole frequency range of the transmitter. For military communications the sinewave must be highly stable and accurate and this is achieved by using a frequency synthesiser, which will be described later.

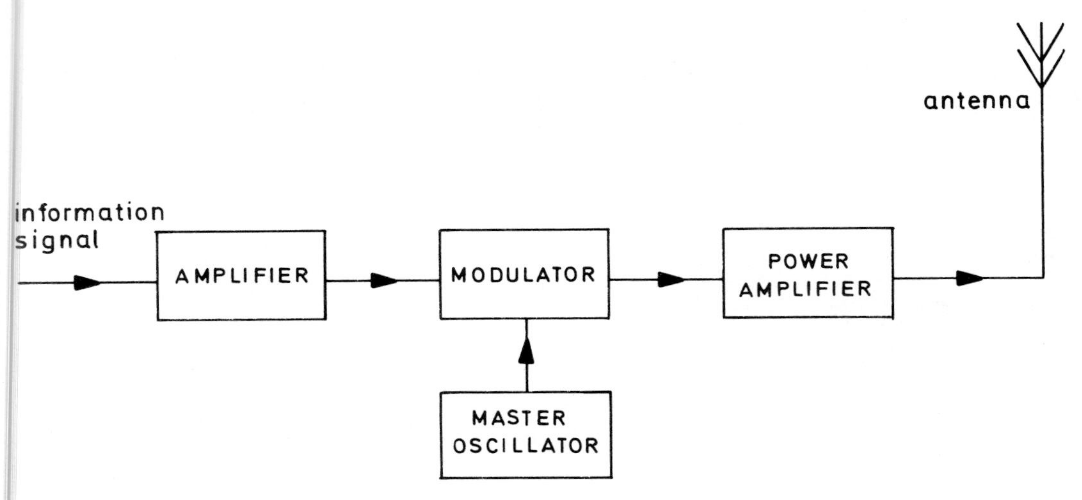

Fig. 3.1 Basic transmitter

b. Modulator. In this unit the information signal is superimposed on the output signal of the oscillator to produce the necessary modulated wave. This may be one of the many types of modulation previously described in the section on modulation in Chapter 2.

c. Power Amplifier. This amplifier generates the necessary power to be fed to the antenna system via a suitable feeder such as a coaxial cable.

d. Antenna. The type of antenna depends on usage; for an all informed net, an omnidirectional antenna is usual so that all round coverage is obtained.

The main performance features of transmitters are:

a. Frequency stability. Usually quoted in parts per million (p. p.m.) over the specified temperature range of the set; using modern frequency synthesisers the stability is usually in the order of a few parts per million.

b. Power. This is dependent on use, for example a manpack set uses only a few watts, a vehicular borne set uses tens of watts and a broadcast station requires kilowatts of power. Radios are also extremely inefficient and a set may well radiate less than 20% of the power it consumes.

c. Spurious outputs. The spectrum of the transmitted signal must be 'clean' which means that it should not contain frequency components other than that of the carrier and the modulation side bands. A 'dirty' spectrum gives rise to electromagnetic compatibility (EMC) problems in that other electronic equipment is affected by this undesirable radiation. Spurious outputs are reduced by filtering and are quoted in dB's relative to the carrier level, their typical order being 40-60 dB's below the carrier level.

RECEIVERS

A basic receiver was shown in Fig. 2.19 to illustrate the process of demodulation, but the more complicated superheterodyne receiver shown in Fig. 3.2 is by far the most common form for high quality communications.

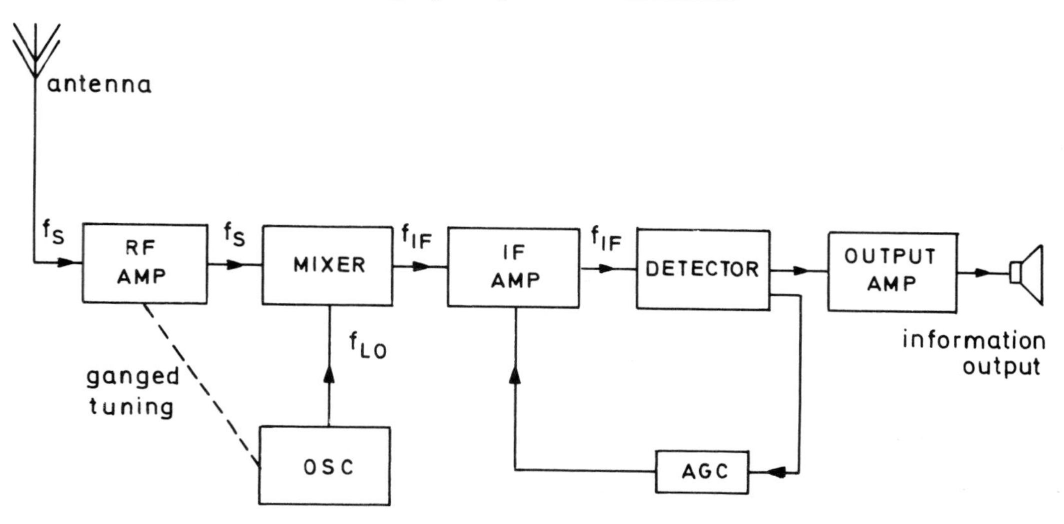

Fig. 3.2 Superheterodyne receiver

The modules are:

a. Radio frequency (RF) amplifier. This amplifier increases the power of the weak incoming signal. The filtering of the amplifier combined with the antenna tuning, selects the wanted signal and rejects unwanted frequencies. Any noise, generated within this stage, that lies in the passband of the tuning is treated in the same way as the wanted signal and is subject to the full amplification of the following stages. Thus the RF amplifier must generate as little noise as possible.

b. Local oscillator. The output from this stage is a sinewave of frequency f_{LO} usually greater than that of the RF signal f_S such that

$$f_{LO} - f_S = f_{IF}$$

The tuning mechanism of the oscillator is ganged to the tuning of the RF amplifier stage so that a constant frequency difference, the intermediate frequency f_{IF}, between the two signals is obtained. The frequency stability characteristics of the receiver are obviously just as important as that of the transmitter and for military sets frequency synthesis is the common practice.

c. Mixer. This stage mixes the RF signals with the sinewave from the local oscillator. The mixer is designed so that several frequencies are produced; the original input signals plus the sum and difference frequencies. The output tuned circuit of the mixer selects the intermediate frequency.

d. Intermediate frequency (IF) amplifiers. This stage is usually a set of high gain tuned amplifiers, the bandwidth of which is designed to pass the signal which was obtained from the mixer stage. These amplifiers are fixed tuned and are designed to give good stability characteristics together with the required band-pass response for selectivity.

e. Automatic Gain Control (AGC). For gradual changes in signal strength due to propagation effects, the AGC is designed to maintain a constant output thus removing the need for continuous operation of the volume control.

f. Detector. The detector extracts from the IF signal the original information signal plus any noise that has passed through the preceding stages. The type of detector must obviously match the type of modulation used.

g. Output amplifier. This amplifies the information signal to a suitable level to be fed to whatever terminal equipment is at the receiver output.

The main performance features of a receiver are:

a. Sensitivity. This is the ability of the receiver to amplify small signals to a usable output level. It is usually quoted in terms of a few microvolts at the RF input producing a certain signal to noise ratio at the output. A typical value is 1 μV input to produce 10 dB S/N output.

b. Selectivity. The tuned circuit band pass filters of the RF and IF stages should reject all but the wanted signal.

c. Fidelity. If the modulated signal within the RF and IF stages is distorted then the fidelity of the output suffers. This distortion can be caused by too narrow bandpass filters or non-uniform amplification.

d. Spurious frequency responses. Non-linearities in the receiver amplifiers and the various mixer products can result in unwanted signals entering the receiver passband. Once this has occurred they cannot be removed by filtering. The technical specification of a receiver gives the levels of adjacent channels, image channel, intermodulation and cross modulation signals. The problem of these spurious signals is described in the following section.

SIGNALS IN A SUPERHETERODYNE RECEIVER

Adjacent Channel

Figure 3.3 shows the two channels either side of the wanted signal. These adjacent channels are rejected by the bandpass filter of the IF so that after mixing the IF signal contains only the wanted channel, shifted in frequency.

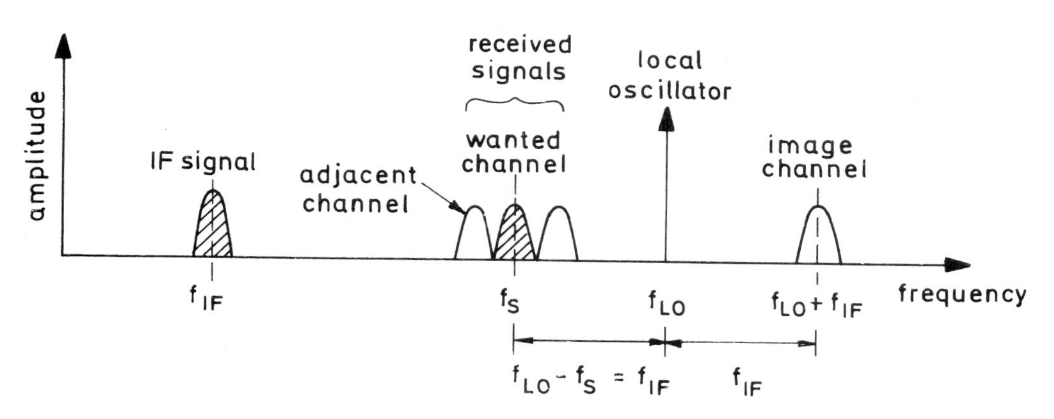

Fig. 3. 3 Signals in a superheterodyne receiver

Image Channel

If the radio receives an unwanted signal which happens to be at $f_{LO} + f_{IF}$ then it will mix with the local oscillator to give an output at the intermediate frequency. Once inside the IF amplifier this signal cannot be removed. Thus it must be prevented from reaching the mixer by the filtering of the RF amplifier. This can be achieved by using as high an intermediate frequency as possible so that the image channel is as far from the wanted passband as possible, as can be seen in Fig. 3. 3.

Intermodulation

As well as the image channel, many other sum and difference frequencies of multiples of input signals and the local oscillator are produced. In general the RF amplifier filters out these intermodulation products except when the level of the signal is very high. This situation can occur if a receiver is placed too near to the transmitter of another set.

Cross Modulation

If a large unwanted modulated signal almost saturates an amplifier then its modulation varies the gain. Hence this modulation will be impressed onto a smaller wanted signal. Although subsequent filtering can remove the large signal its modulation has distorted the wanted signal. This effect is known as cross modulation.

FREQUENCY SYNTHESISERS

A large number of frequencies can be synthesised from a single highly accurate and stable master oscillator. The synthesiser output and the master oscillator are divided by integer values to nominally the same frequency and compared.

This can be simply accomplished by digital electronic circuits. The synthesiser output is continuously adjusted to cancel out any errors that are detected.

The tuning of a synthesiser is not continuous but the steps can be made small enough to select all the channels within a frequency band. The frequency spectrum of a synthesiser is extremely clean as the output is generated directly, although locked to a master oscillator, rather than by building up from several sources.

For a military standard radio, a synthesiser is used as the local oscillator of both the transmitter and receiver sections. In a radio net the transceivers will all be drifting independently from their nominal carrier frequency and so good stability is required.

HF NET RADIO

Military nets use radios in two frequency bands, HF and VHF. HF sets have a long range capability but the band is narrow and highly congested. Secure speech is not currently possible at HF because encryption techniques depend on the signals being digitised and this process requires considerable bandwidth as explained in the Digital Techniques section of Chapter 4. However for low bit rate methods such as telegraphy, secure communications are possible.

Line of sight propagation can be achieved throughout the HF band. In addition a surface wave component can extend coverage beyond the horizon. The losses with this component depend on the type of ground and frequency but in the lower half of the band it is the dominant propagation mechanism. At higher frequencies any surface wave is rapidly attenuated and propagation is by a direct spacewave; thus communication is limited to line of sight only. Although obstacles produce shadow effects some communication is still possible provided the obstacles are small compared with the wavelength of the radiation.

As described in the propagation section of Chapter 2 the ionosphere can reflect signals at HF when the working frequency is less than the maximum usable frequency. Long range communications of more than 100 km, can be established with a transmitter power of a few tens of watts. There are difficulties in using this skywave mode of transmission because the layers of the ionosphere vary seasonally, daily and hourly. These variations are plotted on HF skywave prediction charts which are used to aid frequency planning. Frequencies must be changed often to maintain good communications.

At night the lowest ionospheric layers are weak and signals can be reflected by the higher exposed layers allowing long distance communication. However this process also introduces interfering signals from a long distance. The practical effect of this is that fewer interference free channels are available for any communication within the HF band at night. The lower frequencies of the HF band are most congested because for skywave the skip distance is reduced: this is the distance below which skywave communications is not possible.

The type of modulation used for speech at HF is usually single-side band as it uses the minimum bandwidth in this congested spectrum and it is an efficient type of modulation.

Figure 3.4 shows a CLANSMAN HF manpack radio. It should be noted that much of the size is associated with making the set rugged and with accommodating power supplies. Rechargeable or disposable batteries can be used and are located in the lower part of the equipment. Providing fully charged batteries to sets in forward locations can pose severe logistic problems. The size and layout of the controls of a military radio must allow their adjustment in modern protective clothing and under adverse conditions.

Fig. 3.4 CLANSMAN HF radio

VHF NET RADIO

At VHF there is no significant surface wave component because its attenuation at these frequencies is very high. Communication is by spacewave and is line-of-sight so that the maximum range is horizon limited. The spacewave has direct and ground reflected components which reinforce and cancel as the range varies. A small movement of the antenna at a static site may considerably enhance the received signal. A mobile experiences signal fading as it moves through the pattern of signal maxima and minima. In a built up area the pattern is complicated by further reflections from buildings.

For long distance working or communications over difficult terrain, rebroadcast stations are employed. Figure 3.5 shows a typical installation which could be

Fig. 3.5 VHF net rebroadcast station

installed in a landrover. The right hand set is tuned to f_1 and receives an incoming signal on that frequency. The signal is demodulated in the normal manner and the output is passed via the interconnecting box to the left hand set. The left hand transmitter is tuned to f_2 and the transmitted signal is modulated by the demodulated output of the other set. Rebroadcast only takes place when signals are received otherwise the station is non-radiating.

There are several important points worth noting.

a. Two frequencies are required for a rebroadcast station.

b. The choice of these frequencies must be such that the radiation on one frequency does not interfere with the reception on the other. This consideration achieves good electromagnetic compatibility (EMC).

c. On large formation nets it is quite normal to deploy 3 or 4 rebroadcast stations to give the area coverage required. As each rebroadcast station needs a different secondary frequency the associated frequency planning problems are much increased.

d. Automatic rebroadcast stations recognise and are switched on by a special tone within the incoming signals.

e. If a jamming signal successfully triggers a rebroadcast station then the effective jamming range is increased.

f. If the rebroadcast station is working on a secure net, then the crew at the station must have full decryption facilities if they need to monitor the net traffic.

As a battle is seldom static the problem of maintaining an all-informed capability over hilly countryside can make rebroadcast siting a difficult problem. Good VHF communication sites use high ground, this obviously increases the range to the radio horizon, and improves the all-round coverage of the antenna. Such sites, however, unless screened by further high ground from the enemy can increase the chances of enemy interception and the jamming of friendly nets. It should be remembered that the best possible radio site for area coverage will seldom be the best tactical site and a compromise will almost always have to be made.

The type of modulation used for VHF net radio is FM. An important feature is its capture-effect in that the largest received signal suppresses other signals, thus effectively capturing the radio. In consequence the boundary between the service area of two transmitters is clearly defined. (This contrasts with AM where interference from other areas increases gradually). These well defined service areas allow frequencies to be reused by other nets elsewhere in the battlefield.

At VHF sufficient bandwidth is available for speech security equipment, which requires more bandwidth than unprotected voice transmissions. Security is a strong point in favour of using VHF as opposed to HF as far as commanders and staff are concerned. Unfortunately the extra size, weight and power consumption restricts much current secure radio to vehicle mounted use only. However manpack versions will undoubtedly become increasingly available in the near future.

TRANSMISSION MODES

Different users of net radio can have different requirements and in consequence a variety of transmission modes should be accommodated by a single radio set.

Staff officers will undoubtably wish to have a voice communications facility. However, there may be times when such transmissions especially at HF, cannot be understood because of interference. On such occasions it is very useful to have a morse facility because a good operator will be able to communicate on a net when other forms of transmission are unworkable.

Other facilities that can be provided, usually via special adaptors will allow telegraph, possibly secure, or other forms of data to be transmitted over radio nets by teleprinter or data terminals. A whole series of small digital devices, usually microprocessor controlled, are now available. They can be coupled into the radio to automatically send, receive and store data messages so that sets may be left unattended for a time. A variety of additional facilities are often provided such as burst transmission and automatic coding and decoding of morse.

TWO-FREQUENCY SIMPLEX

The radio system that has been long favoured by the police, fire and ambulance services and other operators of mobile radio stations in a civilian environment is known as Two-Frequency Simplex and is usually at VHF or UHF.

The basis of the system is a talk-through base station positioned on a high feature or building through which all mobile stations must work, as shown in Fig. 3.6. The main similarity with conventional net radio rebroadcast is that although two

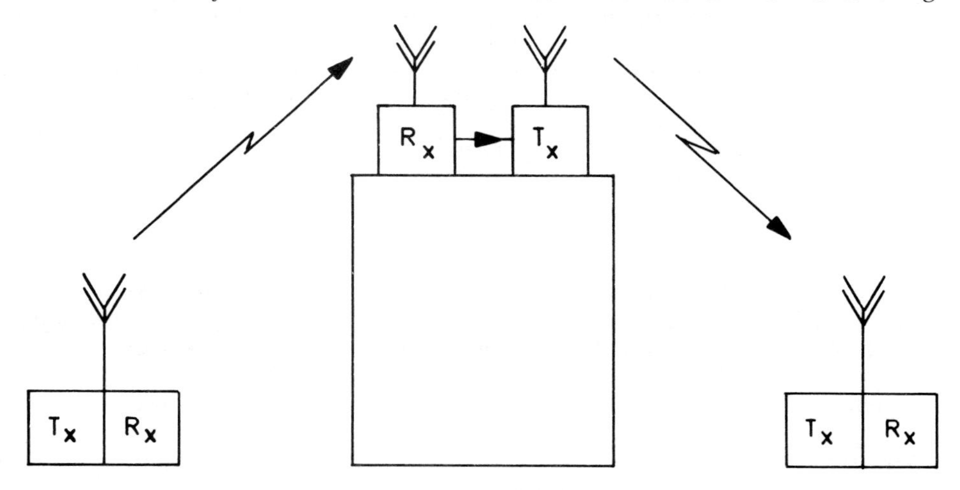

Fig. 3.6 Principle of talk-through

frequencies are used, the system is still simplex. However the important difference is that direct communication between two outstations is not possible as transmission and reception are on different frequencies. The outstations are small, lightweight and have a number of preset channels. Thus they are inconspicuous, do not inhibit the user and are simple to use. For some operations such sets have been bought immediately and require little modification to satisfy military needs. However some attention needs to be given to security and voice procedures as similar sets are easily obtained by the enemy.

The talk-through base station, if well sited, can give excellent coverage in both rural and urban areas. These stations operate automatically and can be left unattended after setting-up. As all outstations use the talk-through its reliability must be high and so a duplicate station is also set-up as a standby.

In general it has proved possible to engineer two-frequency simplex nets as high quality, reliable, all-informed systems. It is however possible for outstations to fail to receive transmissions especially in an urban environment; this is because the many possible signal paths give rise to destructive interference. This multipath effect will also give rise to fading for mobile outstations but usually a move of only a few yards is required to restore reception.

The main drawback of the system in a military context, is the total reliance on the continuous operation of a static central base station.

FREQUENCY PLANNING

Frequency planning for net radio is all important in a crowded radio environment. All transmitters are sources of interference and uncontrolled use of frequencies will lead to very inefficient use of the radio spectrum. Unfortunately the problem is more complicated than just allocating different frequencies to different nets. A transmitter radiates not only the wanted carrier and sidebands but also other signals which are harmonically related to the carrier. This is due to nonlinearities in the output power amplifier of the transmitter. Although these can be reduced by filtering they are still present. A transmitter could readily interfere with a nearby receiver if frequency planning was not adhered to strictly. An example of this for a collocated transmitter and receiver is shown in Fig. 3.7; the difference between the second harmonic of the receiver local oscillator and the

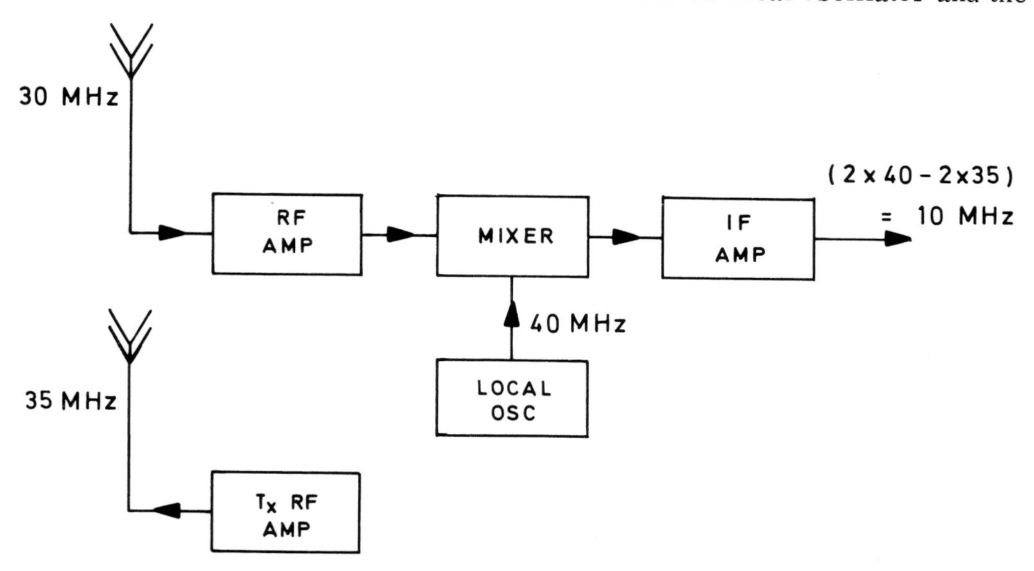

Fig. 3.7 Example of a source of interference

transmitter carrier is equal to the intermediate frequency of the receiver. Thus transmission at 35 MHz can prevent reception at 30 MHz. This is a simple example of the principles that need to be applied when allocating frequencies. In practice it is necessary to consider all transmitters including radars and other radiators not involved directly in communications. The problem is very complicated and difficult to solve.

A powerful transmitter induces very large currents in any metal in the immediate vicinity of the antenna, such as the body of a tank or landrover. Consequently if the vehicle is moving, the metal vibrates and sparks can be seen to arc between separate pieces. These sparks act as transmitters, in fact they are wideband noise transmitters and will certainly interfere with reception.

SUMMARY

Despite its limited capacity, net radio offers an effective means of communicating in a dynamic tactical environment and such systems are likely to remain the primary means of communication forward of divisional headquarters. VHF FM nets are attractive because of their high quality and the possibility of using secure voice, but they are limited to line of sight operations unless rebroadcast stations are used. The rebroadcast stations not only use extra frequencies but are extremely vulnerable to EW attack. HF SSB nets are in general of poorer quality than VHF nets but they do give over-the-horizon coverage without rebroadcast stations. Secure speech, however, because of the limited bandwidths at HF is not generally available.

Although no mention of the EW threat has been made in this chapter, net radio systems are vulnerable targets; this aspect of net radio will be covered in Chapter 5. The future will undoubtedly bring smaller and cheaper radio sets with many new features based on modern technology; details of which are described in Chapter 6.

SELF TEST QUESTIONS

QUESTION 1 At what levels of command is net radio used?

Answer ....

....

QUESTION 2 What factors are important in the design of the radio frequency amplifier of a receiver?

Answer ....

....

QUESTION 3 Why does VHF net radio require good frequency stability?

Answer ....

....

QUESTION 4 What are the relative merits of HF and VHF net radio?

Answer a. HF ....

....

b. VHF ....

....

QUESTION 5 What are the advantages and disadvantages of commercial radio sets that can be bought 'off the shelf'?

Answer a. Advantages ....

....

b. Disadvantages ....

....

QUESTION 6 What steps are taken to ensure that the net radio frequency spectrum is used most effectively?

Answer ....

....

ANSWERS ON PAGE 134

4.
Trunk Communications

INTRODUCTION

Trunk communications link headquarters together and provide the greater part of the communications that formation commanders and staff need. Trunk communications are complementary to the highly mobile and flexible net radio systems used by the fighting troops.

Commanders and staff officers in any one echelon of a headquarters are connected to the headquarters telephone exchange by means of local telephone lines. Calls within the headquarters are made via a telephone exchange which in first generation trunk systems was manually operated.

Calls made from one echelon of a headquarters to another and calls out of headquarters are made via the inter-exchange, or trunk circuits. Since subscribers do not all wish to make a call out of the headquarters at the same time, fewer trunk circuits are required to provide an adequate service. These common user trunks are available to all subscribers within the headquarters. A comparison with net radio shows that trunk systems have a higher capacity for passing information. They provide a more complete service available to the whole staff, with facilities for picture, data and telegraph as well as voice traffic.

Early military trunk communications systems were based on analog transmission, this meant that secure speech was impractical and only telegraph signals could be encrypted. Moreover, the networks had the very serious drawback of following the chain of command, with the communication equipment located at the headquarters, as shown in Fig. 4.1. The consequential role of each headquarters complex as both a tactical base and a large communications focal point or node created serious problems. Firstly, the conflicting requirements of the two roles resulted in serious difficulties in siting the headquarters complex. Secondly, the sheer size of the complex, enhanced these siting difficulties due to the problems of both visual and electronic detection. Moreover, the size and siting difficulties seriously hampered mobility and, without duplication on an impracticable scale, any movement of the headquarters or damage to the network resulted in serious

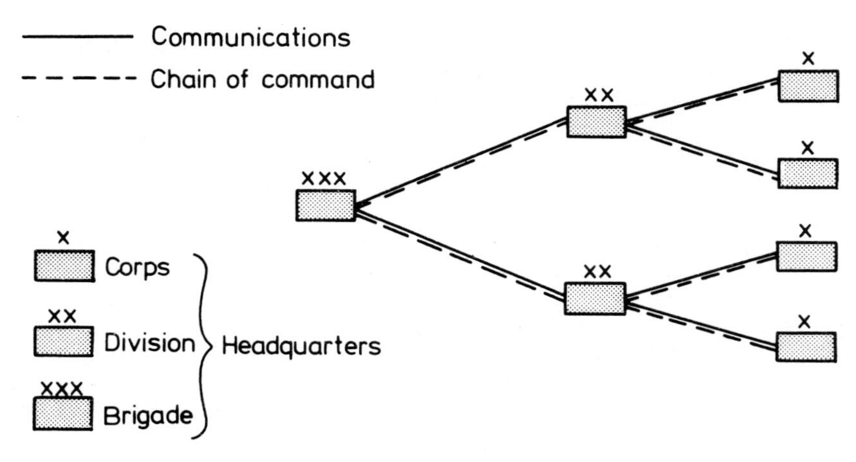

Fig. 4.1 Direct chain of command system

and prolonged disruption of communications. The objective of the second generation British tactical trunk communications system, called BRUIN, was to alleviate these difficulties. The prime factors in achieving this objective are:

a. The use of digital transmission, which permits the encryption of each link to give a secure speech network.

b. The divorcing of the tactical and nodal communication roles of the headquarters complex by the removal of the long haul or trunk radio equipment to a separate communication centre to which the headquarters has access via short links as shown in Fig. 4.2.

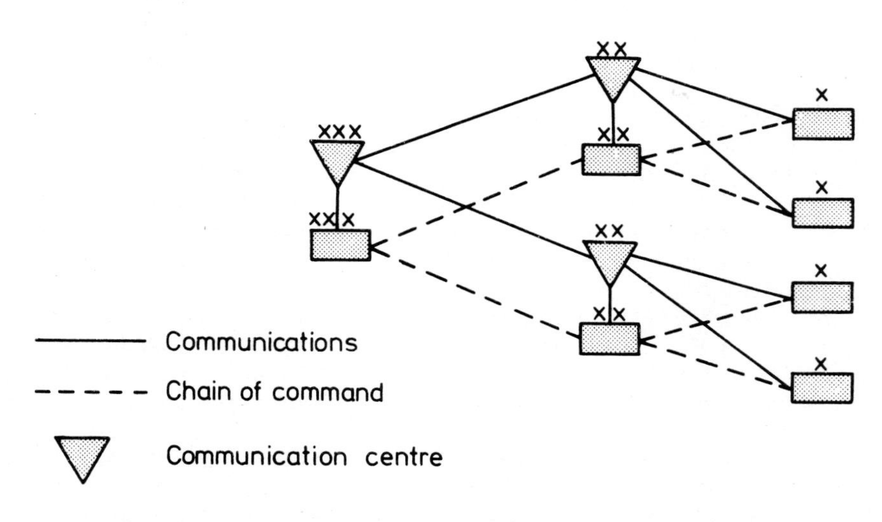

Fig. 4.2 Displaced chain of command system

As a result of this change the communication centres, and not the headquarters complexes, become the nodes of the communication network. The advantages resulting from this configuration are:

a. An appreciable alleviation of the siting problem since both tactical and communication sites can be planned with a higher degree of independence.

b. Easier visual and electronic concealment as a consequence of the smaller physical size and reduction in radio emissions from the headquarters.

BRUIN represents the first phase in the move away from a chain of command system. However, it still follows the chain of command to a degree in that headquarters are only linked directly to communication centres under their own command. An example of this expanded chain of command system is shown in Fig. 4.3

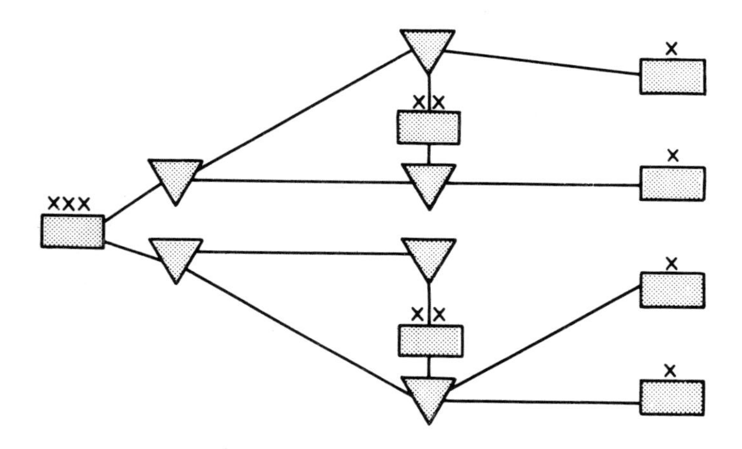

Fig. 4.3 Simple expanded chain of command system

Illustrated is a corps HQ with two divisions each of two brigades. The corps HQ and each division have two communication centres or COMCENS under command. They deploy and move to suit the requirements of their parent corps or division.

There are some additional advantages from this type of deployment. There is a capability to provide duplication on an acceptable scale. This permits mobility without disruption and the headquarters can link to two communication centres. The switching facilities and link duplication provide an alternative routing capability to cover breakdowns, thus improving network reliability.

Within the BRUIN trunk system, formation headquarters achieve movement without loss of continuity by having an alternate headquarters or 'step up' at each level. For example, each corps, division and brigade have two mobile field headquarters complete with their communication head. The communication head or COMHD is a group of vehicles located near the headquarters which provides their local switching and the communications link to the COMCENS for trunk traffic. Figure 4.4 shows that whilst one headquarters holds command the other, the step-up, is free to move to another location and re-establish its communications. When this has been accomplished it is free to accept command and the

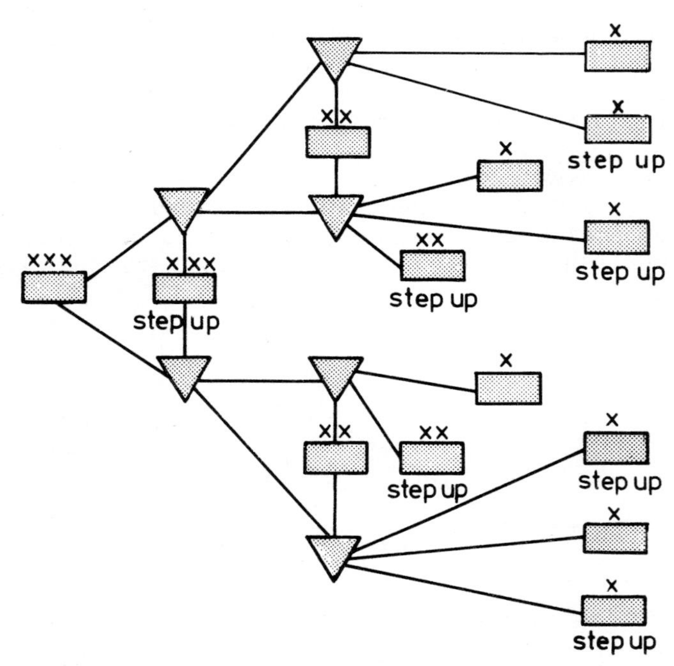

Fig. 4.4 Expanded chain of command system with step-up

other headquarters becomes the step-up and is free to move. This principle of 'one foot on the ground' movement also applies to COMCENS. There are usually two COMCENS in a divisional area. When one of them is required to move, the radio relay links it is working will be passed to the other. For the short time it takes to move and re-establish the free COMCEN, the divisional headquarters and its affiliated brigades will be reduced to one route into the trunk system.

Predictably, the many advantages brought about by BRUIN were not achieved without incurring some penalties. The BRUIN system required more equipment and more sites with a consequent increase in management and maintenance compared to earlier systems. However, these disadvantages were insignificant compared with the improved flexibility, reliability and traffic capacity that the system provides.

The next stage in the evolution of trunk communications systems is the true area system illustrated at Fig. 4.5. The concept is that of a grid of communication centres laid out to give area coverage. Headquarters can move about as required, connecting themselves to the most convenient communications centre. This system can take much more damage before it fails because of the alternative routes available. It also allows much more freedom for movement of the

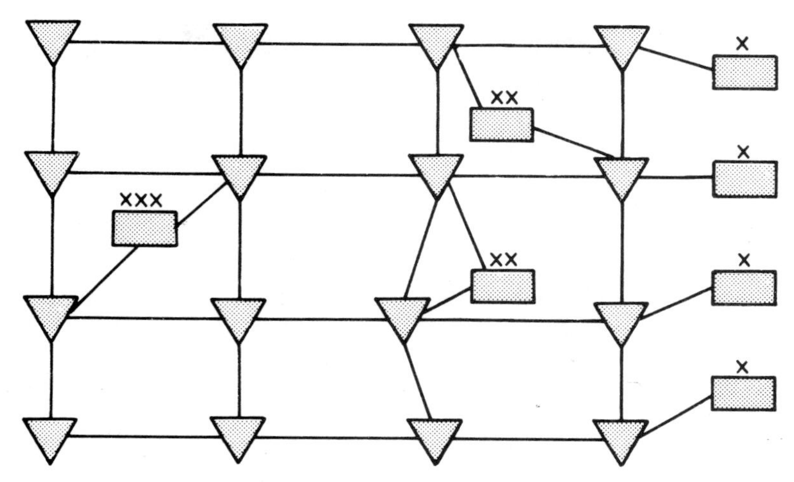

Fig. 4.5 Simple area system

communication centres. This is becoming increasingly important because of modern Electronic Warfare techniques which are described in Chapter 5.

TRUNK SYSTEM FACILITIES

Most tactical military trunk systems have been designed, primarily, to provide telephone links between the staff. These links allow simultaneous two-way (DUPLEX) voice conversation unlike net radio in which the users must switch between transmission and reception (SIMPLEX). However, trunk systems usually support alternative types of communication and may be capable of handling telegraphy, facsimile, television and ADP/computer traffic. The most appropriate of these for a particular message depends on both user and communicator aspects. The staff user is most concerned with the convenience and speed of passing his information. The communicator tries to satisfy these criteria but must also economise on the bandwidth used if he is to provide sufficient channels to satisfy all the staff.

Telegraphy

In telegraphy the message is typed on a teleprinter which converts each letter into an individual 5-bit code word. The trunk system passes the resulting sequence of 0's and 1's to the receiving teleprinter which decodes the sequence and produces a hard copy of the original message. The speed of telegraphy transmission is usually expressed in bauds and for all practical purposes

$$1 \text{ baud} = 1 \text{ bit/sec.}$$

A common rate is 75 bauds which means that the 5 bits of an individual letter's code are transmitted in 5/75 or 1/15th second. For normal English text 75 bauds is equivalent to sending a message at 100 words/minute. Telegraph messages can be prepared on punched tape or typed directly, but for reasonable speed a skilled teleprinter operator is required. Thus telegraphy is a fairly slow method for passing messages but it is very economical on bandwidth, and as will be seen later in this chapter, one telephone channel can be used by several simultaneous telegraph transmissions. Telegraphy is particularly convenient when a message has to be transmitted to several addressees or when detailed information or orders need to be distributed formally.

Telephony

Telephony is the best communication method for conversation because immediate reply is possible. In addition much useful information is conveyed about the identity, personality and mood of a speaker from his speech. These factors are probably of increased importance at the height of a battle when a commander seeks to motivate his forces.

Facsimile

Facsimile is the method by which a copy of a document can be transmitted and it is clearly the best method for sending pictorial information. The document is scanned along a series of lines, similar to the lines on a TV set, and a signal proportional to the lightness or darkness of the document is produced. Military facsimile systems are designed with sufficient resolution or line spacing, to reproduce normal size print in black and white but without shades of grey. The capacity used is that of a single telephone channel and with modern equipment a document of A4 size can be transmitted in about one minute. In the past facsimile has been rather under used because the systems in service took several minutes to transmit a document and even then its quality was poor.

Television

Normal entertainment quality television uses as much bandwidth as a few hundred telephone channels. Thus it is little used for military trunk communication. For the illusion of continuous movement the complete picture must change 25-30 times/second. The bandwidth required can be reduced proportionally to any reduction in this speed and although the resulting picture flickers it may have a useful role in surveillance. Normal speed television is sometimes used within a large headquarters to interconnect staff vehicles, but it requires its own high capacity cable links. Television enables the staff branches to be kept up to date with map briefings without leaving their vehicles.

ADP

The modern trunk system also supports communication between computers. Computer systems transmitting both slow speed data, over a telegraph channel and medium speed data, over a telephone channel, have been fielded. ADP systems have a computer at every headquarters with visual display units available to the staff. The trunk network interconnects these computers so that the same information or data base, is held at each. ADP systems are good for statistical and formatted data such as orders of battle, ammunition returns, casualty returns and stores holdings. As the staff have ready access to the data base the quantity of paper moving within and between headquarters is significantly reduced, as is the demand for telephone channels. For further information on ADP, Volume IX of this series should be consulted.

TRUNK SYSTEMS TECHNIQUES

The engineering steps required in a trunk communication system can be seen in the basic terminal layout of Fig. 4.6. In this chapter these techniques will be

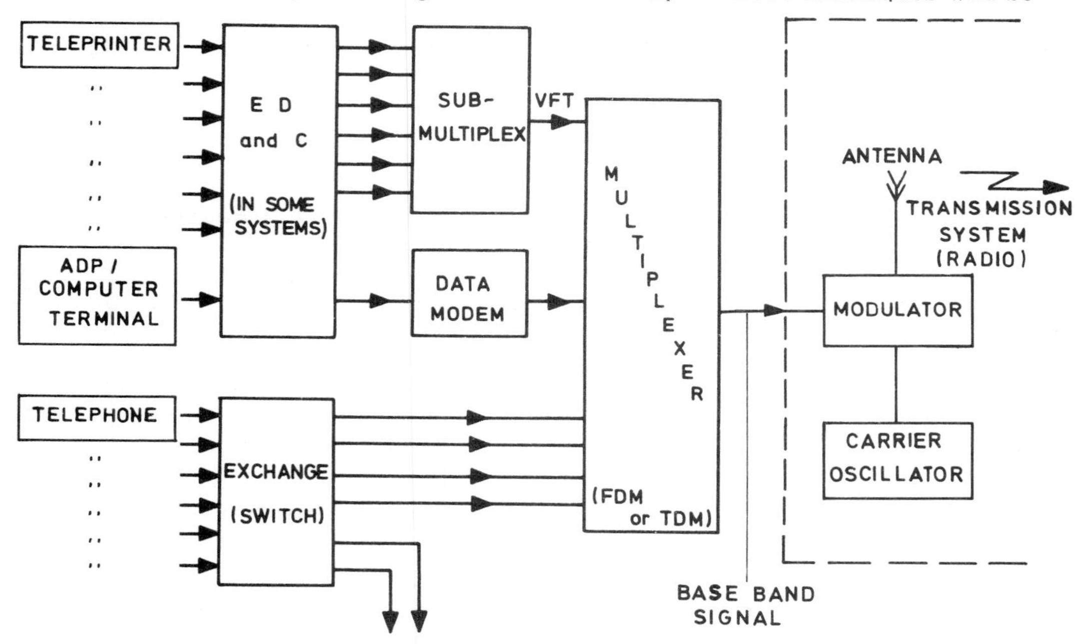

Fig. 4.6 Trunk systems - basic terminal layout

covered in some detail. The basic circuit is the telephone channel and several of these are combined together by a multiplexer. Two alternative techniques frequency division multiplexing (FDM), and time division multiplexing (TDM), will be covered. The combined signal is known as the baseband signal and this is transmitted by, in this case, radio to another terminal where the individual telephone channels are extracted by the reverse process, demultiplexing. Other forms of traffic can use trunk channels. A DATA MODEM matches ADP and computer information to a telephone channel. Whilst because of their slower speed, several telegraphy signals are sometimes multiplexed onto one telephone line using voice frequency telegraphy (VFT). ADP and telegraphy systems often incorporate error detecting codes (EDC), which are used to reduce the number of errors in the received messages due to interference and noise on the radio link. The final area of interest is switching which is required for the messages to reach appropriate destinations.

ANALOG TECHNIQUES

Over short distances such as within a headquarters it may be economical to provide separate cables to each telephone user. For long distances, whether using cable or radio it is essential to adopt multiplexing, so that transmission equipment and channels are shared.

Frequency Division Multiplexing

Single sideband amplitude modulation can be used to transfer speech signals to new frequency bands.

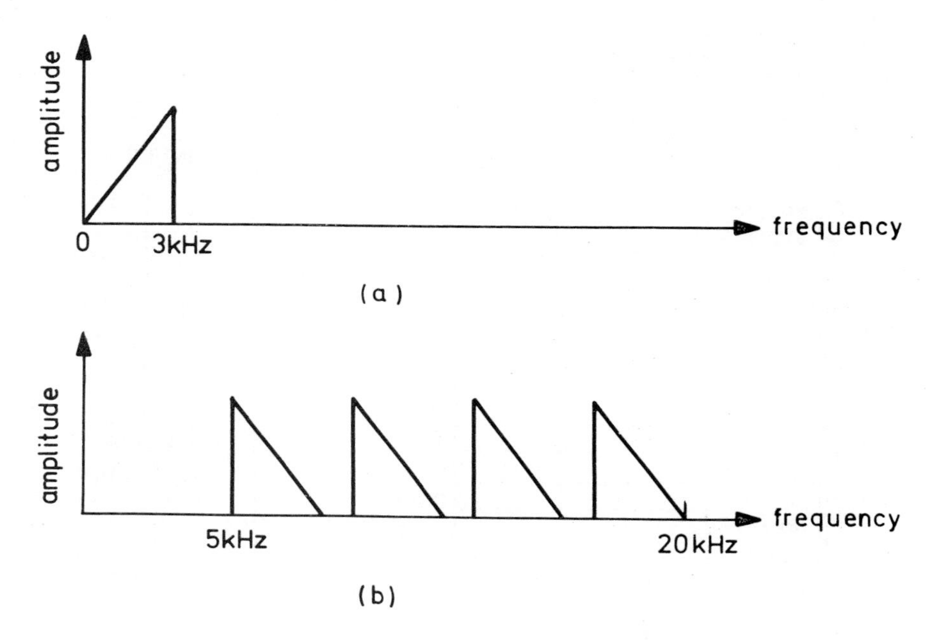

Fig. 4. 7 Frequency division multiplexing

Figure 4. 7a represents a speech signal of 0-3 kHz and Fig. 4. 7b shows four similar signals moved by SSB modulation to share the band from 5 to 20 kHz. The gaps between channels are known as guard spaces and these allow for errors in frequency, inadequate filtering, etc in the engineered system.

Once this new baseband signal, a 'group' of 4 channels, has been formed it is moved around the trunk network as a single unit. A hierarchy can be set up with several channels forming a 'group', several groups a 'supergroup' and several 'supergroups' either a 'mastergroup' or 'hypergroup'. For the numbers of circuits involved in military communications the highest level encountered is the supergroup of between 60-130 channels and these are built up from groups of 15-32 channels.

Groups or supergroups are moved around as single units by the communications equipment and it is not necessary for the radios to know how many channels are involved. A radio can handle a supergroup provided sufficient bandwidth is available. The size of the groups is a compromise as treating each channel individually involves far more equipment because separate filters, modulators and oscillators are required for every channel rather than for each group. However the failure of one module will lose all of the channels associated with a group.

Voice Frequency Telegraphy

The smallest unit of the system is usually the single telephone channel. However traffic such as telegraphy does not require the full capacity which such a channel provides. Thus it is wasteful to use one telephone channel for a single telegraph message typically requiring 150 Hz bandwidth. The technique of frequency division multiplexing can be used to form a group of telegraphy signals such that the group efficiently uses one telephone channel. Thus the effect is for each telegraph signal to be moved to an audio frequency within the voice telephone channel. Hence this technique is known as voice frequency telegraphy.

DIGITAL TECHNIQUES

Over the years various analog processes have been replaced by digital techniques. The military trunk systems currently fielded use a mixture of approaches but their new replacements will be completely digital. In order to understand the advantages of this approach, the principles behind digital techniques will be explained.

In the modulation methods considered in Chapter 2, the signal was used to modulate the amplitude or frequency of a carrier, directly. However in digital modulation a stream of pulses, representing the original signal, is created. This stream is then used to modulate a carrier or alternatively is transmitted directly over a cable. There are two such techniques commonly used in military communication and these are known as pulse code modulation (PCM) and delta modulation.

Pulse Code Modulation

All pulse systems depend on the analog waveform being sampled at regular intervals. The question that must be answered is, how frequently are samples required for the original signal to be reconstructed perfectly?

In Fig. 4.8a it is not unreasonable to expect that from this stream of pulses the shape of the original continuous waveform could be found and fewer samples are really required. Conversely in Fig. 4.8b it would seem unlikely that the original dip, between the two samples shown, would be predicted. It is found that the limit occurs when the waveform is sampled twice in each cycle of the highest frequency present. This case, which is the sampling theorem limit, is shown in Fig. 4.8c. In practice unless a somewhat higher sampling rate is chosen any noise and distortion in the system results in unacceptable degradation in any reconstituted waveform. (Another example of sampling is in the simulation of a continuously moving picture from a series of film frames. The persistence of vision of the eye reconstitutes the original from the limited number of frames or samples.)

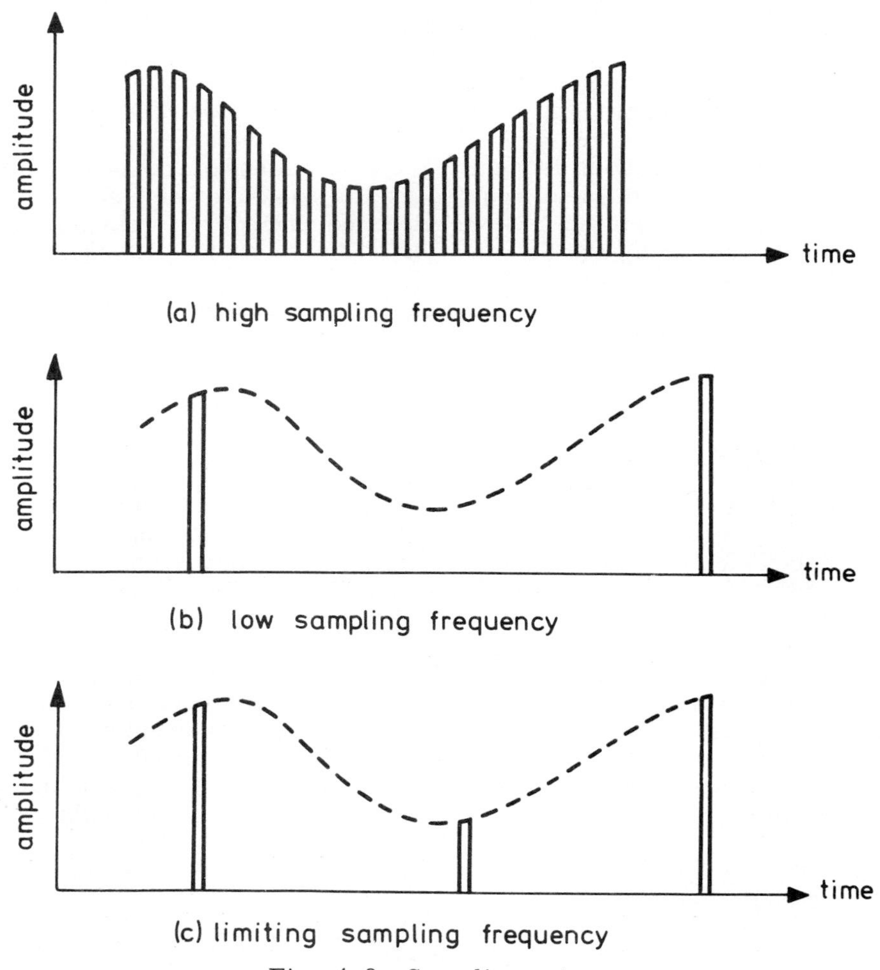

Fig. 4.8 Sampling rate

The signal created by sampling our analog speech input is known as pulse amplitude modulation. It is not very useful in practice but is used as an intermediate stage towards forming a PCM signal. It will be seen later that most of the advantages of digital modulation come from the transmitted pulses having two levels only, this being known as a binary system. In PCM the height of each sample is converted into a binary number.

Figure 4.9 shows the steps required to form a PCM signal. In this example the amplitude of the signal, at a sampling instant, can be represented by any whole number between 0 and 16, the quantising levels. Thus any sample can be rewritten as a 4-bit binary number and it is these numbers which are transmitted

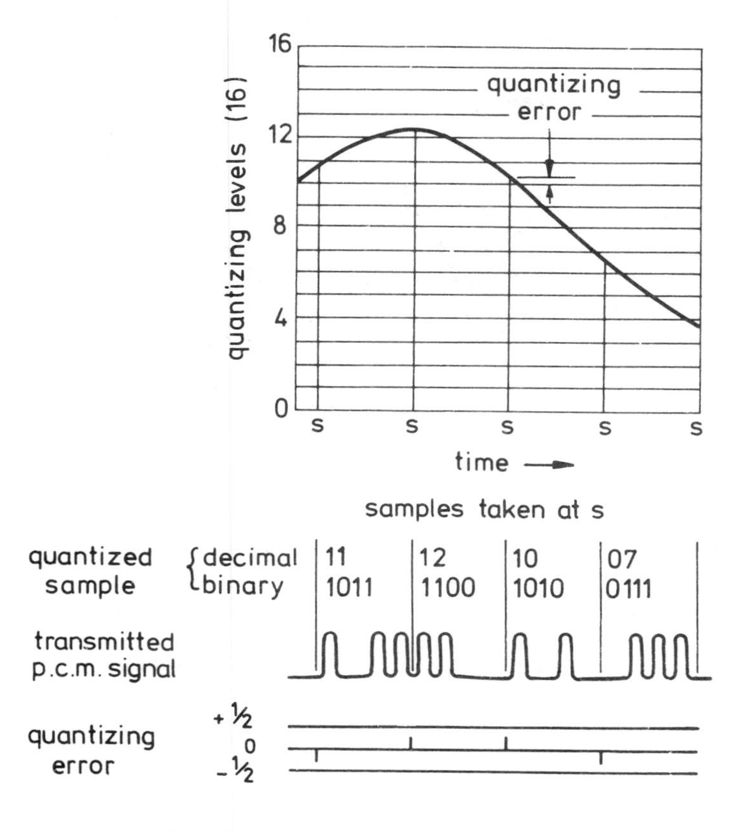

Fig. 4.9 Pulse code modulation

as the PCM digital pulse stream. In this example if it were desired to transmit 3 kHz speech then samples would be required at a rate of at least 6 kHz (ie 6 kbits/sec pulse amplitude modulation). As a 4-bit binary number is required for each sample a PCM pulse stream of 24 kbits/sec would be formed. This example illustrates the point that PCM signals require a greater capacity or bandwidth for their transmission than the original analog waveform.

Returning to Fig. 4.9 it can be seen that the third sample is transmitted as the binary equivalent of 10 although the original signal was slightly greater than this level. Such errors are known as quantisation noise because their overall effect is to perturb the re-constituted signal in a noise like fashion around the correct waveform. The quantisation noise can be reduced by adding more levels. In this example if 64 levels were used rather than 16 for the same input signal range, then the quantising errors would be reduced to a quarter of their original size. However each sample would then require a 6-bit binary number (64 levels) instead of 4-bits (16 levels) and the rate of the pulse stream representing a 3 kHz analog signal would increase to 36 kbits/sec. Thus the bandwidth required for PCM depends on the number of bits in the binary number at each sample and a subjective decision is required on how many levels are adequate for quantisation noise to be acceptably low. The BRUIN trunk links use 64 level PCM.

Delta Modulation

This technique was originally devised as a simpler and consequently cheaper method than PCM. However now that advances in digital electronics have reduced the costs of PCM to a comparable level, delta modulation has been found to give adequate quality speech at lower bit rates. Although it must be appreciated that the bandwidth used is still much greater than the original analog speech.

In delta modulation instead of coding the absolute level of a sample it is the difference between successive samples which is transmitted. Rather than using a multiple bit binary number for this, a single bit only is used to indicate that the original waveform is either rising (1) or falling (0), as shown in Fig. 4.10.

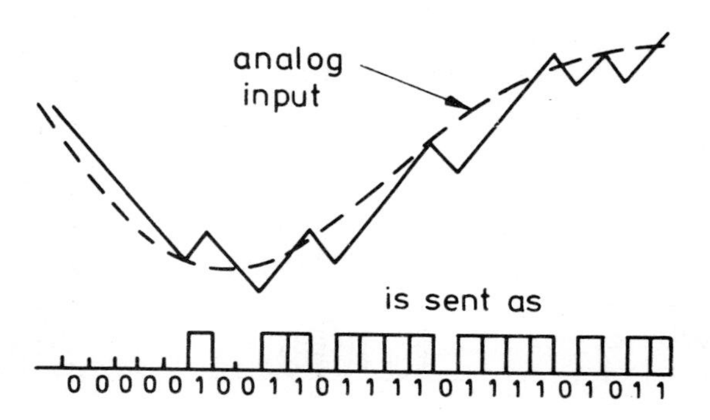

Fig. 4.10 Delta modulation

At the receiver the pulses are used to cause the waveform to rise or fall producing the 'zig-zag' waveform shown. However if this is smoothed out by a filter then the re-formed signal resembles the original analog input.

Time Division Multiplexing

It is possible, with pulse modulation systems, to use the time between samples to transmit signals from other circuits. The technique is known as time division multiplexing, (TDM). To do this it is necessary to employ synchronised switches at each end of the communication link to enable samples to be transmitted in turn, from each of several circuits. Thus several subscribers appear to use the link simultaneously. Although each user only has periodic short time slots, the original analog signals between samples can be reconstituted at the receiver.

DIGITAL MODULATION ADVANTAGES

So far we have seen that signals can be converted into digital form and can share communications links by multiplexing. However the bandwidth required is large and it can only be justified if there are advantages compared with an analog FDM system.

Regenerative Repeaters

A problem with analog transmission is that the cables, radio links, etc distort the signal because noise is added, the different frequency components are attenuated by varying amounts and delays are not constant. The overall result is that there is a transmission distance beyond which the signal becomes unacceptable.

Digital transmissions suffer from the same distorting mechanisms but at the receiver there is only one simple decision to be made, that is whether the pulse is a 0 or a 1. Hence considerable distortion of the signal is permissible before any errors are made in determining the state of a pulse. Providing regenerative repeaters are inserted before wrong decisions are made it is possible to maintain the sequence of 0's and 1's perfectly no matter how long the communications link. Figure 4.11 shows the regeneration process. The weak signal at b is amplified and corrected for any known distortions. Based on the amplified waveform at c correct decisions are made and the new perfect pulses are transmitted from d.

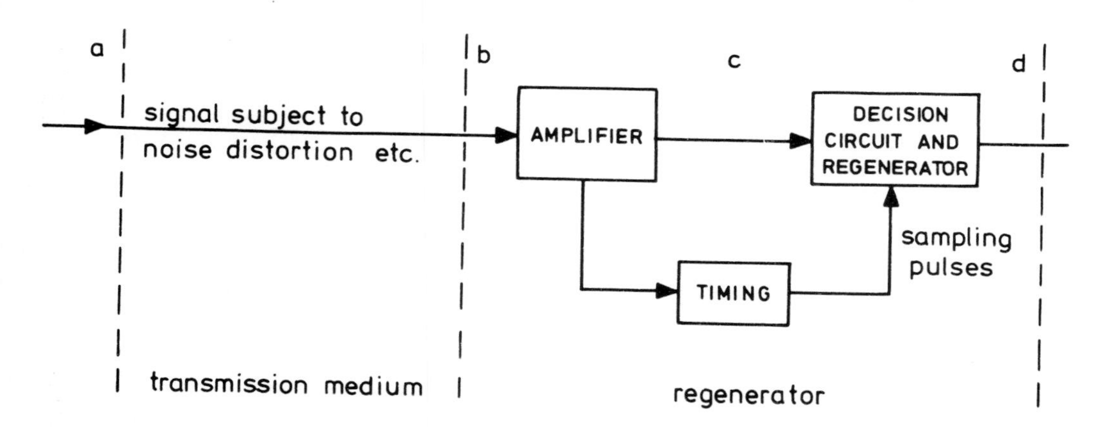

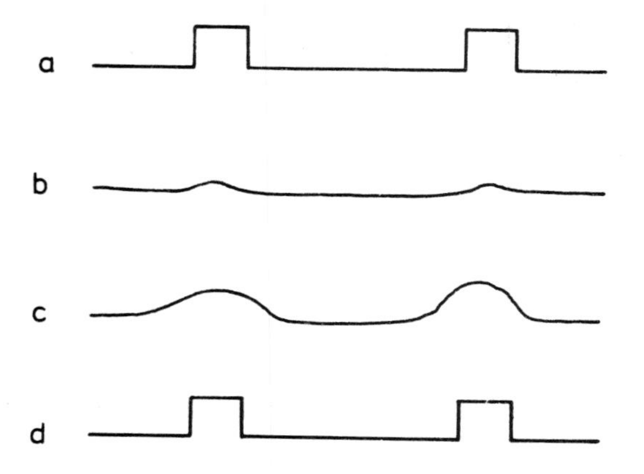

Fig. 4.11 Regenerative repeater

Encryption

An important military advantage of digital modulation is that the signals can be made secure. With analog signals it is possible to scramble the transmission in various ways. Splitting the signal up into frequency bands and rearranging these is possible, but for practical systems there are insufficient combinations for adequate security. Alternatively the signal can be split into short time blocks and rearranging is possible, but this is again too limited for high level security in systems requiring conversation at normal speeds.

The encryption of digital traffic is both simple and effective. At the transmitter a 'random' sequence of pulses is added to form the secure message, whilst at the receiver decoding is achieved by subtracting the same sequence. Such sequences are easy to generate electronically. A truly random sequence cannot be used because the receiver would be unable to generate an identical sequence. Thus a pseudo-random code is generated. This appears to be random to an observer as the pattern of 0's and 1's will not repeat for an extremely long time, in practice this can be several days. However at the transmitter and receiver the electronics generates this sequence from a comparatively simple secure key code.

RADIO RELAY

It is clear that multiplexed groups of trunk signals require wideband communication links. Single pairs of wires can only be used for quite small groups. Coaxial cables, waveguides and optical fibres have sufficient bandwidth for trunk traffic but the expense and time of laying these and their inflexibility usually makes radio preferable.

For military trunk communications, radio relay is used and Fig. 4.12 shows its principle. The location of the terminals makes it impossible to obtain an adequate direct radio path and an intermediate relay station is inserted in the link, hence the name radio relay. This is sited so that there is a good radio path to both terminals and the signals received from one terminal are re-transmitted automatically to the other. In a long link there may be several relay stations.

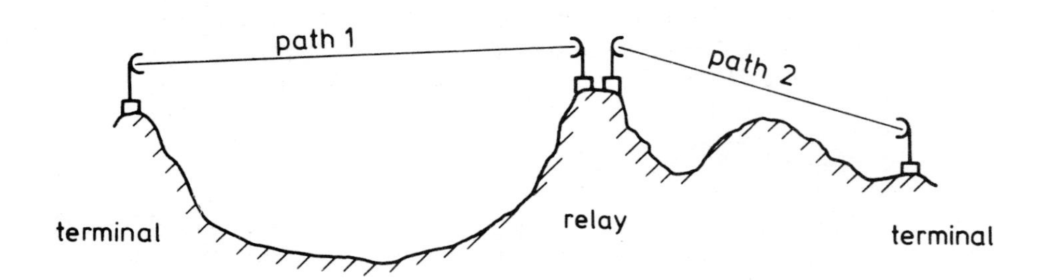

Fig. 4.12 Principle of radio relay

Frequency

Frequencies from 60 kHz upwards have been used for radio telephony, and much telephone traffic has been carried, both by HF skywave, for transoceanic circuits; and by VHF, for short range, near line of sight conditions such as links to land vehicles or aircraft and ship-to-shore. In general these frequencies are inevitably limited in the channel capacity they can provide, and multi-channel equipment is not encountered below VHF.

For civil systems, high capacity is the main objective, and this can best be achieved at SHF, above say, 3000 MHz. 12000 MHz is about the upper limit due to an increasing inability to penetrate rain and other precipitation. SHF systems are limited to line of sight. Sites must be carefully chosen and plenty of time allowed for aligning the equipment on site. This is acceptable for static civil systems. High towers may be employed, with large, highly directional antennas in order to minimise both mutual interference, and transmitter power.

Tactical mobility, and short time into and out of action are requirements which mitigate against use of SHF for Army systems. Instead UHF is favoured as it gives adequate channel capacity, with much more tolerance on siting.

Radio Path Capability

Each link of a radio relay chain can only tolerate a certain loss between the transmitter and the receiver. In a TDM system if this transmission path loss (TPL) is exceeded then the number of errors, in the received digital pulse stream, becomes unacceptably high. After allowing for antenna gain and feeder losses the remaining loss which can be tolerated in the radio link is the radio path capability. Workable radio shots may either be long and unobstructed or much shorter if hills intrude near to the direct line of sight as shown in path 2 of Fig. 4.12.

Antennas

As the system works from point to point, directional antennas, radiating in a specified direction, can be used. Concentrating radio power in one direction only, makes best use of the power. Whilst little radiation in unwanted directions both reduces the interference to neighbouring radio systems and minimises the possibility of interception. On reception directivity minimises the interference from other stations, including hostile jammers. Typical antennas for UHF trunk links were shown in Figs. 2.26 and 2.27. Their gain is restricted to about 10-25dB so that alignment and concealment are straightforward.

Siting

Radio relay terminals situated on high ground will give the best possible radio path. Unfortunately, for obvious tactical reasons headquarters are usually sited in more concealed positions. Thus the headquarters must be connected to a radio relay site by a 'tail', which is a cable or subsidiary radio relay link.

From a map reconnaissance, possible radio relay sites can be picked out. Then by plotting path profiles it is possible to confirm whether or not the resultant paths have sufficient clearance of obstacles taking into account the desirable Fresnel clearance which was discussed in Chapter 2. In practice the final choice of sites involves a reconnaissance on the ground to check the following:

a. Local obstacles not shown on the map, in particular tree height and density.

b. Possibilities for concealment and camouflage.

c. Access for the radio relay vehicles.

d. Whether other users of electronic equipment have already occupied the site because radar and surveillance devices require similar high features.

Repeaters

Figure 4.13 shows a radio relay repeater. The weak incoming signal, at frequency f_1, is converted down to a lower intermediate frequency where it is

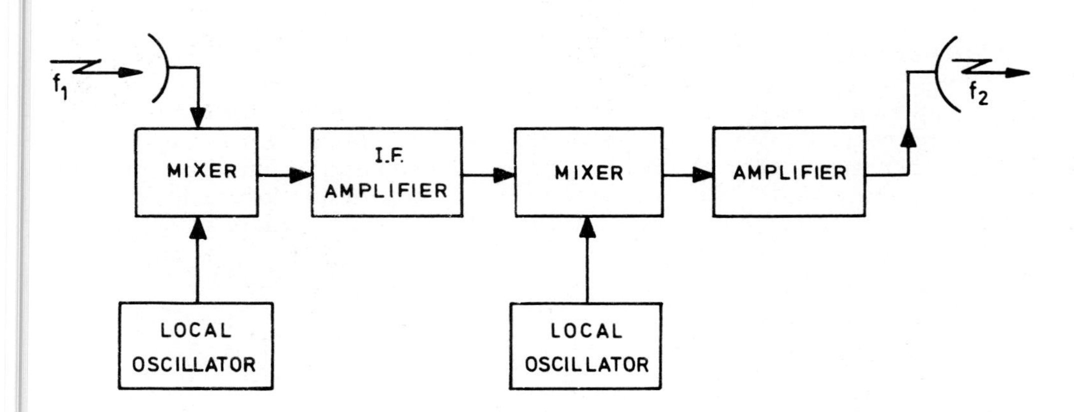

Fig. 4.13 Radio relay repeater

amplified as in the superheterodyne receiver described in Chapter 3. The signal is then converted up to a suitable frequency, f_2, for onward transmission over the next radio link. Although the frequencies f_1 and f_2 are in the same band and experience the same propagation phenomena there is always a frequency difference between the input and output of a repeater. If this were not the case, then wanted signals coming into the receiver would be swamped by any signal leaking back from the transmitter output, and the whole repeater could become unstable.

In fact the choice of frequencies is quite complicated. Electromagnetic compatibility considerations mean that radios colocated at the same physical site must not use the same frequencies or any multiples, or other combinations related to the local oscillator and intermediate frequencies within the sets. It should be realised that all radios, whether used for net or trunk communications or telemetry and command purposes increase this EMC problem. In addition, using repeaters increases the number of frequencies required and a frequency allocation plan is needed for the entire trunk network, including periodic changes to counter enemy electronic warfare efforts and movement of friendly forces. Over a large network frequencies can be re-used in widely separated areas.

The TRIFFID radio is shown in Fig. 4.14. The lower unit is the power-supply.

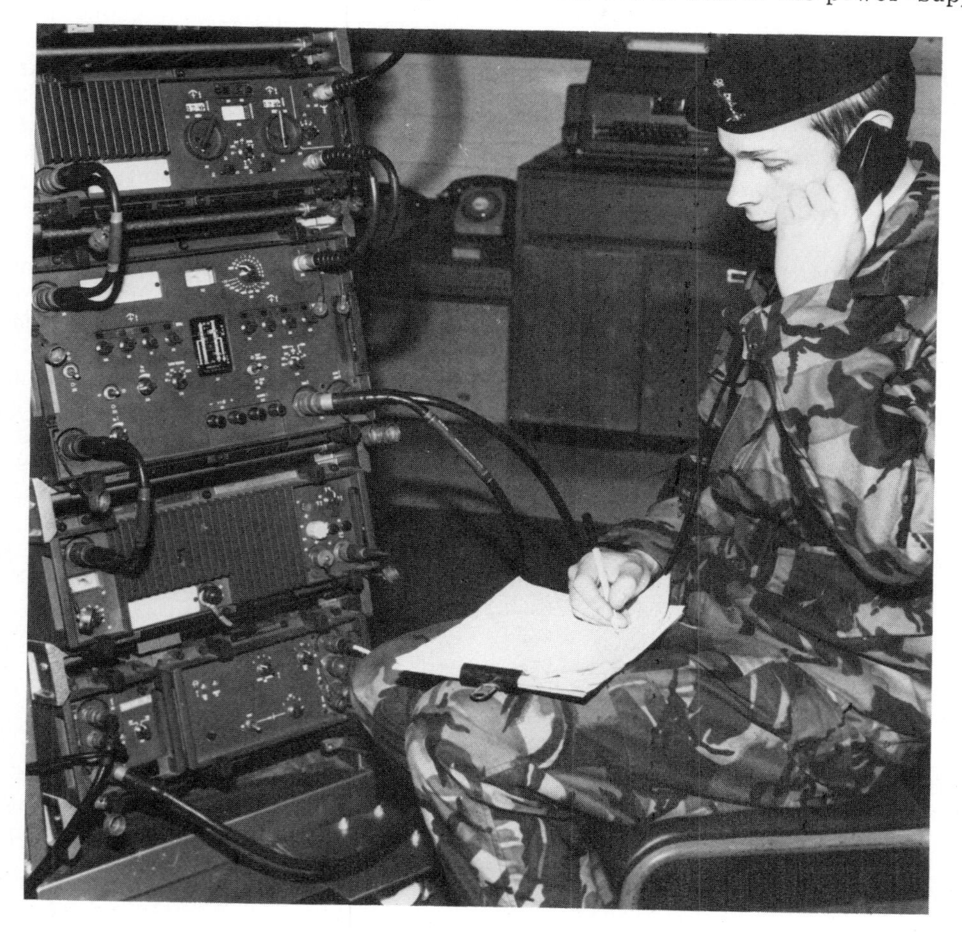

Fig. 4.14 TRIFFID radio

The middle, a systems unit which contains the baseband circuitry, modulators, demodulators and frequency synthesisers. The upper unit is one of three alternative radio frequency heads covering frequencies between 225 and 1850 MHz.

TRIFFID and other modern trunk radios contain built in test equipment, BITE. Continuous monitoring of its own performance provides visible and audible alarms on the systems unit. In addition, to ensure the immediate pin-pointing of a fault in an established radio relay chain the set can be made to introduce pattern generator and recognition circuitry. By the operator setting simple switches the signal can be made to 'loop' round appropriate parts of the chain to rapidly identify the trouble area.

TELEPHONY SWITCHING

A permanent dedicated channel between two remote terminals is only economic if there is a need for almost continual communication. The more usual situation is for a terminal to be able to communicate with any number of other terminals at unpredictable times. This is achieved by connecting the terminals to a network which possesses a switching capability. The terminals are then interconnected by whichever links are appropriate. Although switching can be performed manually, automatic systems now dominate and this section describes different approaches.

What matters most to a user is the reliability of the system. The traffic engineer expresses this by the term 'Grade of Service', which is the proportion of attempted calls which fail to be connected due to under provision of equipment. A Grade of Service of 0.01 means that one call in every hundred will not be connected. The service will be satisfactory if sufficient trunk links and exchange equipment are provided. Over provision increases costs, whilst under provision causes considerable frustration during the 'busy hour', thus traffic forecasting is of great importance. Standard formulae relate the amount of equipment required to the amount of traffic for a given Grade of Service. In civil systems these calculations are based on the traffic intensity in the busy hour of the day, which is mid-morning. The traffic intensity is the average number of calls simultaneously in progress. The statistics of military traffic are less clear and the busy hour is related to events rather than the time of day.

Analog Telephone Switches

Various forms of electro-mechanical switches have been used. The Strowger type was invented in 1891 but is still the most common type of switch in the UK civil system and is used reliably and successfully in BRUIN. In most countries Crossbar switches are more common and their form leads more directly to the modern electronic controlled crosspoint switches, as used in the German AUTOKO Military Trunk System.

All three systems effectively close contacts in a switching matrix or space-switch as shown in Fig. 4.15. It can be seen that two calls are in progress between input 2 and output 3 and input 4 and output 1.

Both Strowger and Crossbar require involved electro-mechanical devices which are only cheap because of the large volumes manufactured. Faults increase with age and when contacts become worn and dirty, and their speed is limited. In modern electronic exchanges the reed relay shown in Fig. 4.16 gives improvements in speed, cost and reliability.

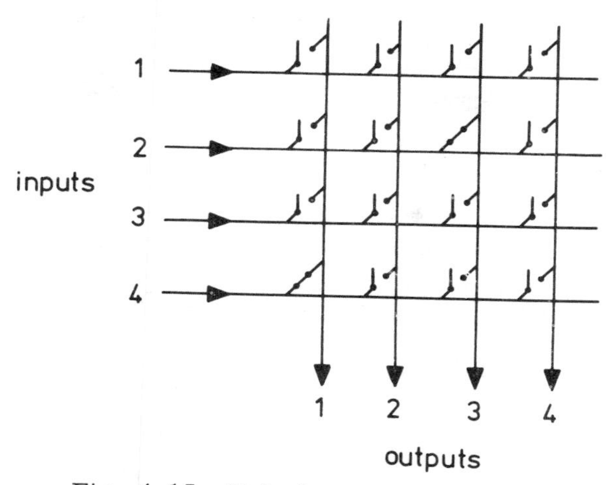

Fig. 4.15 Telephone space-switch

When a current is passed through the coil the reeds become magnetised and are attracted to give electrical contact. In the absence of a current the tips are separated by about 0.1 mm. Due to this very small movement the switch can be operated much more rapidly than the older types. The sealed glass tube contains nitrogen and the contacts remain clean.

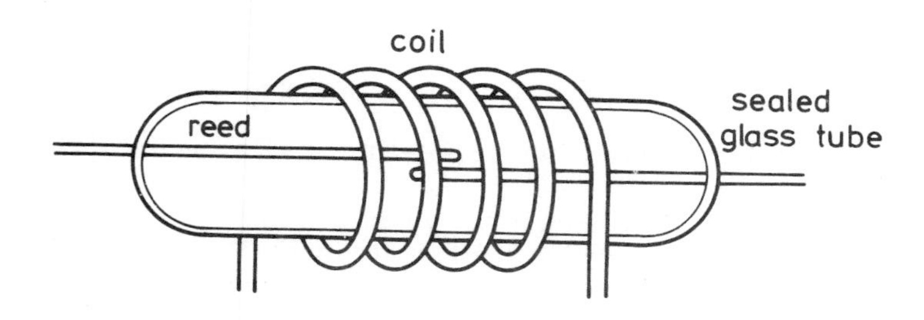

Fig. 4.16 Reed relay switch

The three systems are all suitable for analog traffic and the reed relay cross-point may be computer controlled in an 'electronic exchange'. The advantages of computer control will be described later. However high speed digital signals cannot be handled successfully because switch contact bounce and surges of current in the exchange can lead to false pulses.

Digital Telephone Switches

A true electronic switching element, suitable for analog signals, has not been found. This is because unlike a mechanical switch the on/off resistance range of a semiconductor based device is limited, and signals coming out of these devices are not faithful linear copies of their inputs. Thus analog signals distort and leak into nominally off outlets, this effect is seen as increased background noise. However although unsuitable for analog traffic these switches are good enough for binary digital signals as even when distorted, decisions on whether the signal state is a logical 0 or 1, can be made.

Time-Shared Digital Crosspoints

Digital transmission links are connected to a digital exchange in time division multiplexed form. To connect one channel from an input group to an output group requires two operations as shown in Fig. 4.17.

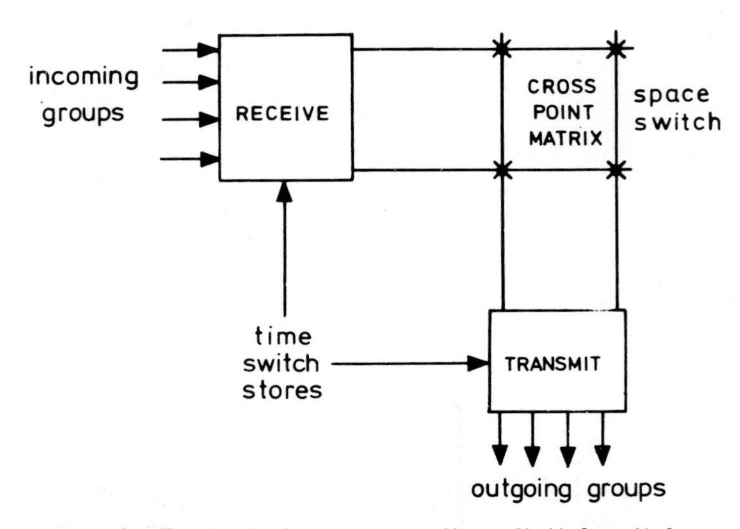

Fig. 4.17 Basic time-space-time digital switch

First it is shifted from its time slot in the input group to the one it will occupy in the output group, this being known as time switching. Secondly it must be switched through from the input system to the output system through a crosspoint matrix, this being known as space switching. The space switching can be carried out by electronic crosspoints, probably electronic logic gates, operated for just a few microseconds in each time slot. They are thus shared between a number of simultaneous calls.

The PTARMIGAN switch is of this type, the traffic in time division multiplexed as a 32 channel group and the input signals are stored until the appropriate output time slot for switching through the space switch. The switch is controlled by a minicomputer.

Software

The activities a computer carries out depends on its software, the program changes can be made by rewriting a few lines of program and alterations to the hardwiring may be unnecessary. This gives a computer considerable flexibility even after it is deployed and provided sufficient spare capacity is available new tasks may be added. Volume IX of the series expands on this subject.

It is essential that software is of good quality and well documented and procedures have been adopted to achieve this. A system such as PTARMIGAN required more than 50 man years of programming and without a well ordered approach, 'bugs' would be extremely difficult to find. The quality of software is a concept which it is hard to measure. Basically the program must enable the system to carry out the required task and cater for every contingency. The program is not unique and there are trade offs between the space it occupies within the computer, the speed of performing tasks, ease of checking and self diagnostics. In addition the 'ultimate' program may take a prohibitive number of man hours to develop.

In pre-digital types of telephone switching catastrophic failure of an exchange was rare. However a poor computer program could 'crash' (or fail) with only a small fault. Thus there are dangers in relying on software for telecommunications switching. Adequate reliability levels are achieved by the software initiating alternative actions if individual components, say, fail. Some systems carry out frequent diagnostic checks on hardware and reconfigure to prevent total failure.

The principal advantage of a software approach is that features and facilities of the telephone system can be modified by reprogramming. The hard-wired Strowger and Crossbar systems require a complete rebuild to modify the facilities provided.

Computer control can readily provide the following features:

a. Precedence system; automatically disconnecting a lower precedence subscriber.

b. Conference and broadcast calls; predetermined or set up by the subscriber.

c. Alternative routing to reduce congestion or avoid any failed links.

d. Standard directory with compressed and abbreviated dialing.

TELEGRAPHY SWITCHING

The pattern of telegraph traffic is generally different to telephony because there are far fewer teleprinters and each is used more frequently than a telephone subset. In addition many of the peaks of traffic are smoothed because messages are waiting to be typed on the teleprinter. Thus with different traffic statistics an alternative type of switching network is appropriate. Other differences from telephone traffic are that a 2-way link is not always required, messages may be multi-addressed and delays in transmission may be acceptable.

Through Switching

The technique of interconnecting teleprinters in the same manner as telephones is known as through switching. It is used in commercial TELEX. Separate links and switches of a higher quality than telephone channels are required because errors are more damaging to telegraphy than to speech.

Tape Relay

All messages are prepared and stored on punched paper tapes. At a relay, incoming messages are punched on to a new tape by a reperforator. They are then torn off for processing or onward transmission. Hence it is often known as a 'torn tape' system. The routing is carried out manually, the operator takes the tape and identifies its destination address. The tape then awaits its turn for transmission from the appropriate autosender to the next relay centre. Tape relay is more economical on the number of channels than through switching, as messages can queue up when a route is heavily loaded.

Automatic Store and Forward

At high traffic levels automation of the switching becomes economical. The basic form of a store and forward switch is shown in Fig. 4.18. In torn tape the routing was manual and the messages were stored as paper tapes. The first stage of

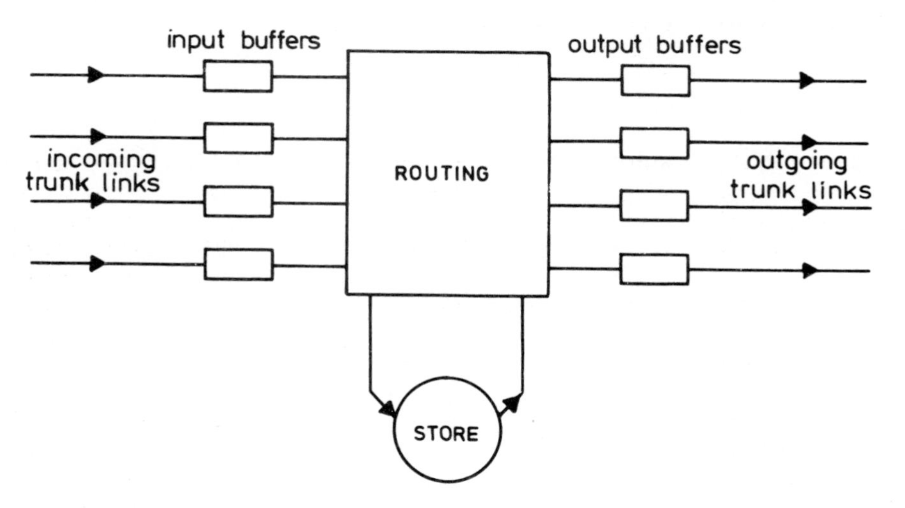

Fig. 4.18 Store and forward switch

automation is to reduce the amount of paper tape: in a busy military field vehicle the quantity can become unmanageable. This can be achieved by punching out only the address, at the start of the message, and storing the rest of the message on magnetic tape. The human operator can cope with certain errors in the routing information of the signal. Further automation requires very few errors in the routing information. Error detecting and correcting codes are often incorporated in this information to counteract errors during transmission.

Completely automatic store and forward systems have been used successfully, one is the British Telegraph Automatic Routing In the Field (TARIF). A computer completely controls the message routing with overall monitoring and supervision from a control console.

MODERN TRUNK SYSTEMS

Behind the newest generation of military trunk systems lie several basic principles which distinguish them from earlier types.

Convergence

In the older systems specific channels could handle one communication method only. For example one channel may have been used for telegraphy, probably with VFT, one channel to carry an ADP computer based system and another dedicated exclusively to a priority telephone user. The problem then is that all the other users are sharing the remaining channels and the grade of service becomes unacceptable. The idea of convergence is that all channels can handle all types of traffic and none are dedicated. It is still possible to have a priority user so long as one channel is left but the particular channel to be used can vary. With users sharing a larger number of channels there is statistically a better chance of making a successful call and the Grade of Service rises.

Commonality

Commonality arises directly from the principle of convergence. It is the use of identical devices or system blocks for as many different roles as possible. For example a digital system can accept telephony, telegraphy, facsimile and ADP data and convert each into a bit stream. Only one type of switch is required to route any of these around the network. Commonality reduces the costs of basic modules, particularly in electronics where the unit costs fall rapidly with volume of production. The reduction in types of modules naturally simplifies the problems of stocking spares and servicing faulty items.

Commonality is used in 'hardware' items such as the electronic circuits and it is also used for 'software' ie the computer instructions. This is helpful in the management and operation of these large complicated computer based systems.

Digital Systems

Many trunk systems have gradually replaced analog techniques by their digital equivalents, but the full advantages of digitisation are not achieved until this process is total. For example if a pulse stream, from a trunk link, encounters an analog telephone switch then a digital-analog conversion must take place. The opposite process takes place for the signal to pass out along another digital link. After several of these transformations the signal becomes seriously degraded.

Modern trunk systems are digital throughout. The technical features of the UK PTARMIGAN system highlight this approach. Voice, telegraph, data and facsimile inputs all share the same digital channels (16 kb/s). Speech is digitised directly at the users telephone subset, by delta modulation. Encryption of the traffic takes place within the staff vehicles where the trunk groups, which are 32 channel TDM, are formed. The channels are routed by computer controlled digital switching using a time-space-time method. Only one type of computer is used throughout but alternative software enables it to be used for conventional switching, telegraph store and forward and system management thus achieving a high level of commonality. Thus the advantages of digitisation are fully achieved.

There are other advantages in that moving to a digital system is consistent with the general direction and thrust of both electronic and computer engineering. In electronics we are seeing increasingly complex integrated circuits (ICs), containing thousands of transistors, but they are manufactured in such vast quantities that their prices are falling. The impact of microprocessors and other digital devices is clear in many areas of life today. However this electronic revolution has only occurred in digital electronics and large scale integrated circuits are rare in analog systems.

The original prototype PTARMIGAN subset contained about 200 integrated circuits and was large and expensive. The redesigned subset shown in Fig. 4.19, uses a

Fig. 4.19 PTARMIGAN subset

microprocessor and now contains 30 ICs and three hybrid circuits. The microprocessor replaces all the original hardware except for the analog parts and some high speed logic circuits.

Minicomputers can now be emulated by several microprocessors sharing the tasks, this means that they can be replaced by alternative components which give the same performance at the input and output terminals. Replacing a minicomputer by microprocessors at a communications switch, say, will give a useful improvement in size and in power consumption: a minicomputer requires considerable power for cooling and air conditioning.

There are great similarities between a modern digital trunk network and the organisation of computers, minicomputers and microprocessor based systems. The convergence of telecommunications and computing means that techniques originally evolved in one discipline are now relevant to the other. ADP systems involving the interconnection of several computers fit naturally into computer controlled trunk systems.

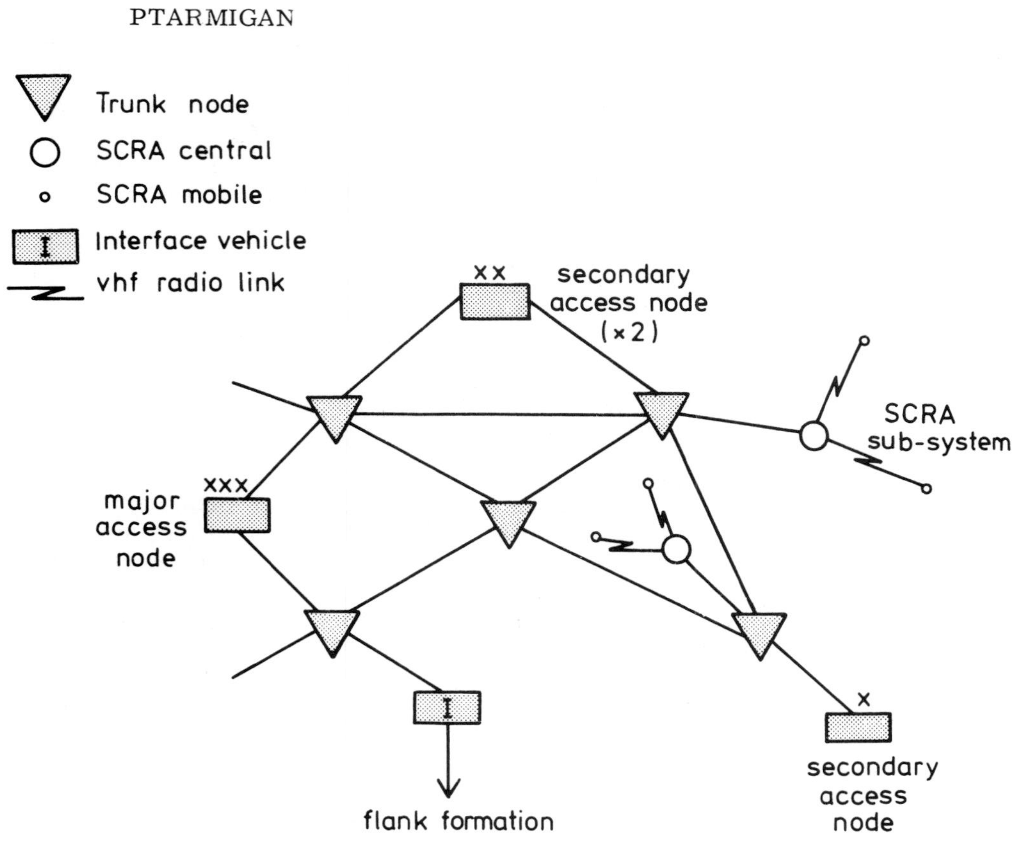

PTARMIGAN

Fig. 4.20 Section of PTARMIGAN trunk system

PTARMIGAN is the newest secure, digital, tactical trunk communications system for use by the British Army and Royal Air Force in North West Europe. Its principle is that of a true area system with a communications network covering the entire area of operations into which headquarters and mobile users connect. Digital techniques and computer controlled switching allow the subscribers, who each have deducible extension numbers associated with their appointment, to be found automatically wherever they are within the system. Another innovation and something entirely new in military communications systems is the extension of full trunk communication facilities to mobile users with the introduction of the Single Channel Radio Access (SCRA) sub-system.

Figure 4.20 on the previous page shows a section of the trunk network and the various types of access to the PTARMIGAN system. These will be explained in the following sections.

Trunk Network

A trunk network is the backbone of the system and comprises a network of nodes (previously known as COMCENS) deployed to cover the area of operation. The nodes contain stored program controlled digital electronic circuit switches which are interconnected by radio relay links. The links are normally less than 25 km long and carry either sixteen or thirty two 16 kb/s channels in time division multiplexed streams. The trunk nodes and radio relay must be static when operating but because of the redundancy built into the system individual nodes are able to move frequently, both to keep pace with tactical movement and for their own protection.

Static Access

Staff in headquarters are allocated their own small communications complex, an access node, which gives radio relay access to the trunk network. The corps headquarters is served by a major access node which also provides local switching for up to 150 subscribers. Staff users' equipments are connected to it by cable or by 'down the hill' SHF multichannel radio relay, if a good radio path is available. The switching facilities provided by a major access node for a large group of staff subscribers such as a corps HQ would be unnecessarily complex for a small group such as a brigade HQ. In this case terminal equipments serving the brigade staff are linked via a secondary access switch and radio relay vehicle directly to the most convenient trunk node where all switching even for local calls, takes place. The secondary access switch can cater for up to 25 subscribers. A divisional HQ is similar but a little larger and uses two secondary access switches.

Isolated or Mobile Access

The isolated or mobile user, not located at a headquarters, uses a type of radiotelephone known as Single Channel Radio Access (SCRA) which is connected to the trunk network through a radio central. This gives the mobile user the same facilities and service as a static user but with complete freedom of movement throughout the entire area by being semi-automatically reconnected to successive radio centrals. A description of mobile subscriber access systems and the possibility of their future use as 'stand alone' systems is included in Chapter 6.

Characteristics of PTARMIGAN

The system is designed and deployed to give many alternative routes between trunk nodes and automatically avoids damaged or faulty equipment and traffic congestion when placing calls. As long as one path is available subscribers can place calls to each other. Two system features ensure this: they are flood search and delegated routing. By sending out a flood search message a switch is able to discover the location of any wanted subscriber as indicated in Fig. 4.21.

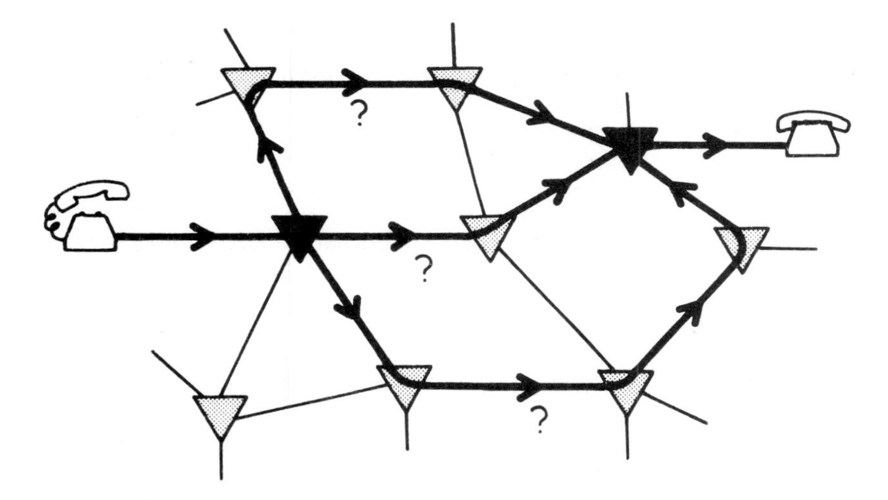

Fig. 4.21 PTARMIGAN flood search

To minimise the number of flood searches which use many paths, a frequently called list FCL, is compiled at each access node. When a subscriber is called his number and location is placed at the top of the FCL. Thus if the same subscriber is called from this access node soon afterwards a flood search is not required. However if a subscriber is not called for some time then he moves down and eventually off the FCL. The other important feature is the delegation of control of routing to each switch in turn in the setting up of a call. This enables a switch to use its local knowledge to avoid damage or congestion and so direct a call to the wanted terminal as shown in Fig. 4.22.

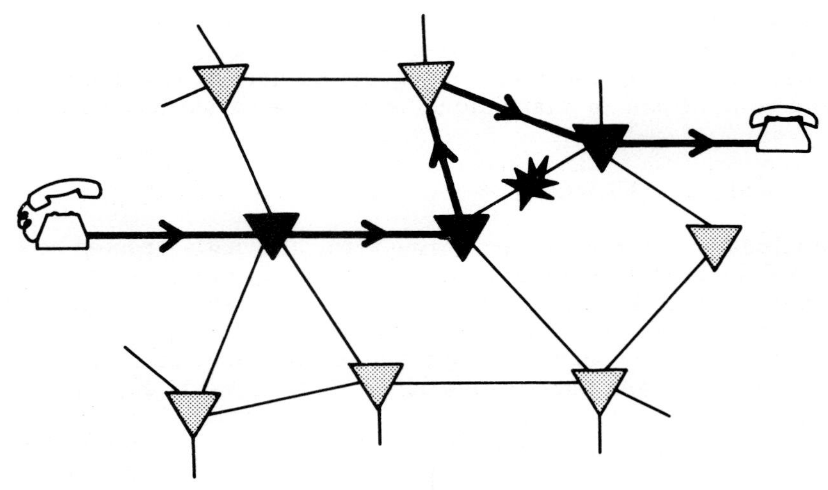

Fig. 4.22 PTARMIGAN automatic alternative routing

Other characteristics are as follows:

Service. The PTARMIGAN network provides a high speed switching system which allows calls to be connected within seconds. The probability of a connection at the first attempt depends on the precedence used but should not be less than 95% for a routine call.

Modes. The system provides secure facilities for voice, telegraph, data and facsimile. The basic mode is voice, but telegraph, data and facsimile equipments can be connected to the subscribers terminal unit via an adaptor unit. The message centre provides telegraph and facsimile facilities for users who do not have their own equipments.

Directory. Every user by virtue of his appointment has a fixed and unique 7-digit number which is deducible and based on service organisation. For example, the first digit represents the formation, the second digit the type of formation, the third the identity of the formation and so on. The computer-controlled switches locate the wanted user wherever he is situated in the system.

Management. Comprehensive system management facilities are provided using the same type of processor as the switch and store and forward, enabling the system to respond quickly to a changing military situation.

Repair. Built in test equipment (BITE) is incorporated in all hardware designs for speedy repair of most faults by module replacement. Difficult faults are cleared in the field using automatic test equipment (ATE) installed in electronic repair vehicles.

Facilities

One of the advantages of a computer controlled electronically switched system is that the facilities that can be offered to subscribers are almost limitless. This is because, generally speaking, each facility will be controlled by a section of computer program or 'software' and not expensive special to task electronic components or 'hardware'. PTARMIGAN facilities include the following:

Compressed dialling. Users who have frequent need to call the same numbers anywhere in the network need to dial only 3 digits, made up from a special directory code of 2 digits followed by a serial number 0-9, representing up to 10 frequently called numbers.

Line grouping. A group of users of common interest such as a staff branch in the same headquarters, are able to have all incoming calls switched to the first free line of the branch in a preferred sequence. The size of the group is limited to 5.

Call transfer. By dialling 5 digits a user is able to cause all his calls to be automatically transferred to another number during periods of temporary absence.

Call forward. Either party is able to transfer an existing call to a third party. For example, a call made to Operations which should have been made to Artillery can be forwarded to Artillery without the need for complete re-dialling.

Call hold. A user is able to leave a connection that has been made and, using the same telephone, make a second call which cannot be overheard by the waiting caller: subsequently he can return to the original connection.

Exclusive user circuits. Certain users require high speed connections to one or more other users on a private line principle. This requirement is catered for on a limited scale, since exclusive user circuits reduce the number of trunk lines available to others. Users with this facility require an additional telephone for normal calls.

Pre-emption. Two degrees of pre-emption are available, PRIORITY and FLASH, selected by a push button. This facility enables subscribers with pre-emption to automatically disconnect calls of lesser importance and be connected. The disconnected caller receives a special tone informing him that he has been overridden by a pre-emptive call. This facility operates throughout the system on both trunk and local calls.

Conference and broadcast. A conference is a multiple connection between 3 or more users. A broadcast is a similar connection although only the originator may transmit. All modes can be used for broadcast. A conference is conducted using voice or teleprinter.

Barred trunk access. This is a minimise facility whereby only vital users can make trunk calls. It is used to reduce the load on the system during periods of critical activity or when it has been damaged. There are 2 stages of trunk barring, each affecting different numbers of users. A barred user is not able to make external trunk calls but may receive incoming trunk calls and make local calls.

Indirect teleprinter messages. Store and forward equipment is located at certain trunk nodes to provide for distribution of teleprinter messages. Store and forward delivers multi-address messages in precedence order and, if necessary, stores a message for an addressee who is temporarily disconnected from the system.

Interfaces with other Systems

PTARMIGAN conforms with internationally agreed EUROCOM standards and will, therefore, interface with other EUROCOM compatible systems. Fundamental to this is the single channel capacity of 16 kb/s which other NATO countries, who subscribe to the EUROCOM agreement on standardisation, will use when introducing their own new communications systems. Tactical interface installations as shown in Fig. 4.20, can be deployed to the flanks and the rear of a PTARMIGAN equipped corps to facilitate communications with flanking corps and to higher command. Such interface installations also connect PTARMIGAN to BRUIN equipped units and formations. Additionally there is a facility, widely available throughout the PTARMIGAN system, to interface with static telephone and telegraph systems, both civilian and military.

Another important facility is the ability of combat net radio users to gain single voice channel access into the PTARMIGAN system. This is achieved by deploying Combat Net Radio Interface (CNRI) equipments to good communications sites throughout the system. All calls across the CNRI are manually set up by the interface operator with the radio subscriber calling in on a designated CNRI frequency.

SELF TEST QUESTIONS

QUESTION 1 What are the relative merits of a chain-of-command and an area system from the point of view of:

a. System control?

Answer ....

....

b. Flexibility of response to changing communication requirements?

Answer ....

....

c. Survivability?

Answer ....

....

d. Communications security?

Answer ....

....

QUESTION 2 Discuss, from the point of view of both the user and the communicator, the advantages and disadvantages of:

a. Telegraphy.

Answer ....

....

b. Telephony.

Answer ....

....

c. Facsimile.

Answer ....

....

d. Television.

Answer ....

....

QUESTION 3 List the advantages and disadvantages of using a digital modulation technique for speech rather than analog transmission.

a. Advantages.

Answer ....

....

....

b. Disadvantages.

Answer ....

....

QUESTION 4 List the types of switches which have been used in trunk communications.

Answer ....

....

QUESTION 5 What facilities should a modern trunk system provide?

Answer ....

....

ANSWERS ON PAGE 135

5.
Communications EW

INTRODUCTION

Electronic Warfare (EW) is a subject of vital concern to all users of electromagnetic equipment on the battlefield. This chapter is an introduction to its concepts and principles. The discussion will be limited to communications EW: EW in other fields, such as radar and electro optics is covered in Volumes VII and VIII in this series.

Recent advances in electronics have increased the versatility and capacity of field communications systems. As commanders and their staff have become accustomed to these much improved facilities their dependence on them has grown. Since a potential enemy will be well aware of this dependence, he can be expected to mount a considerable effort to turn this to his own advantage. In broad terms communications EW is the exploitation or degradation of the enemy's use of the electromagnetic spectrum whilst protecting our own ability to use the same spectrum, as and how we wish.

Structure and Components

The three quite distinct components of EW and the relationships between them are shown in Fig. 5.1 overleaf. They are:

a. Electronic Support Measures (ESM) which include Search, Intercept, Monitoring and Location of enemy transmitters.

b. Electronic Countermeasures (ECM) which seek to prevent or effectively reduce the enemy's capability to use his communications systems by jamming; or to deceive the enemy by electronic means.

c. Electronic Counter Countermeasures (ECCM) which cover the protection of friendly communications from the enemy's interception, deception, jamming and location efforts.

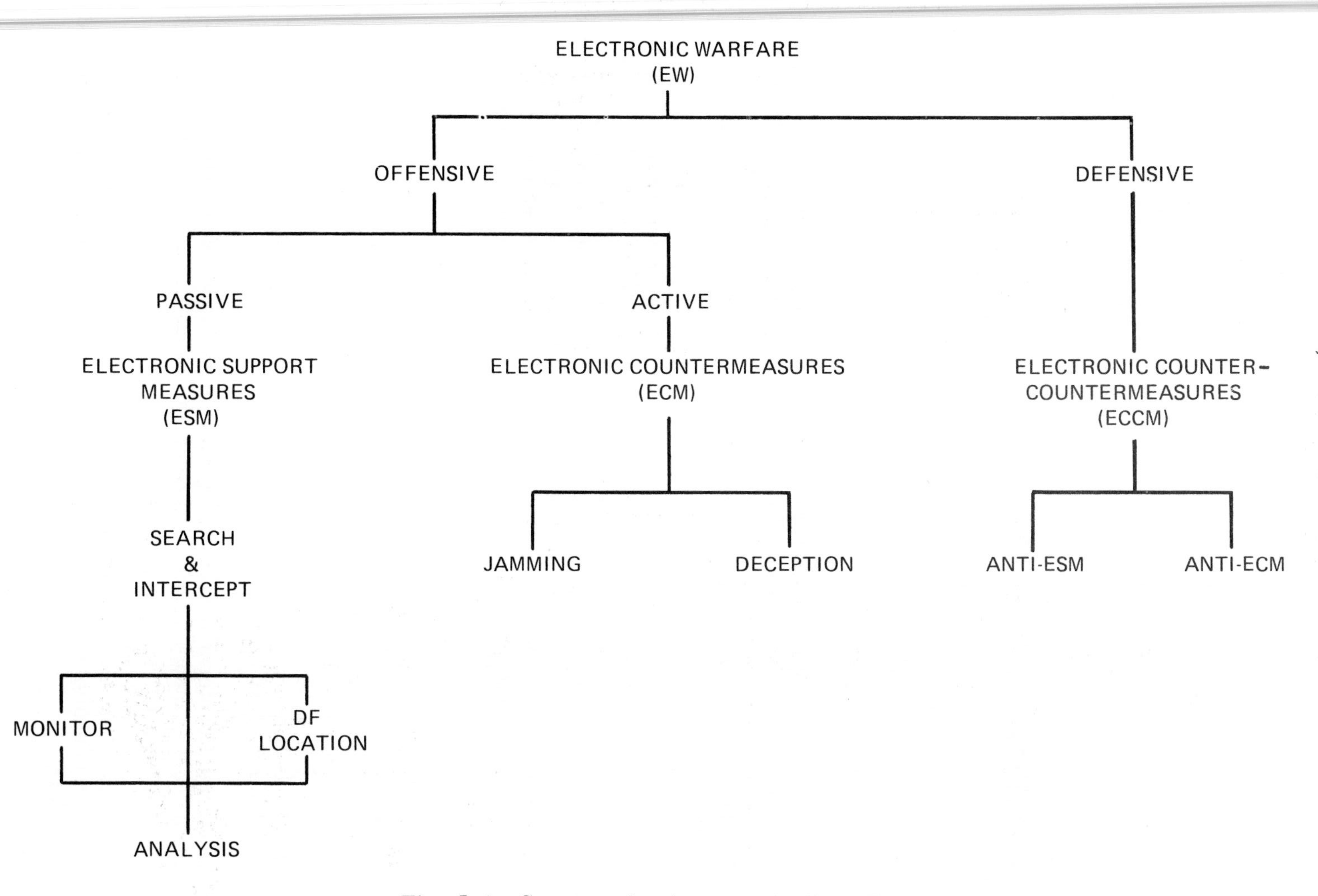

Fig. 5.1 Components of communications EW

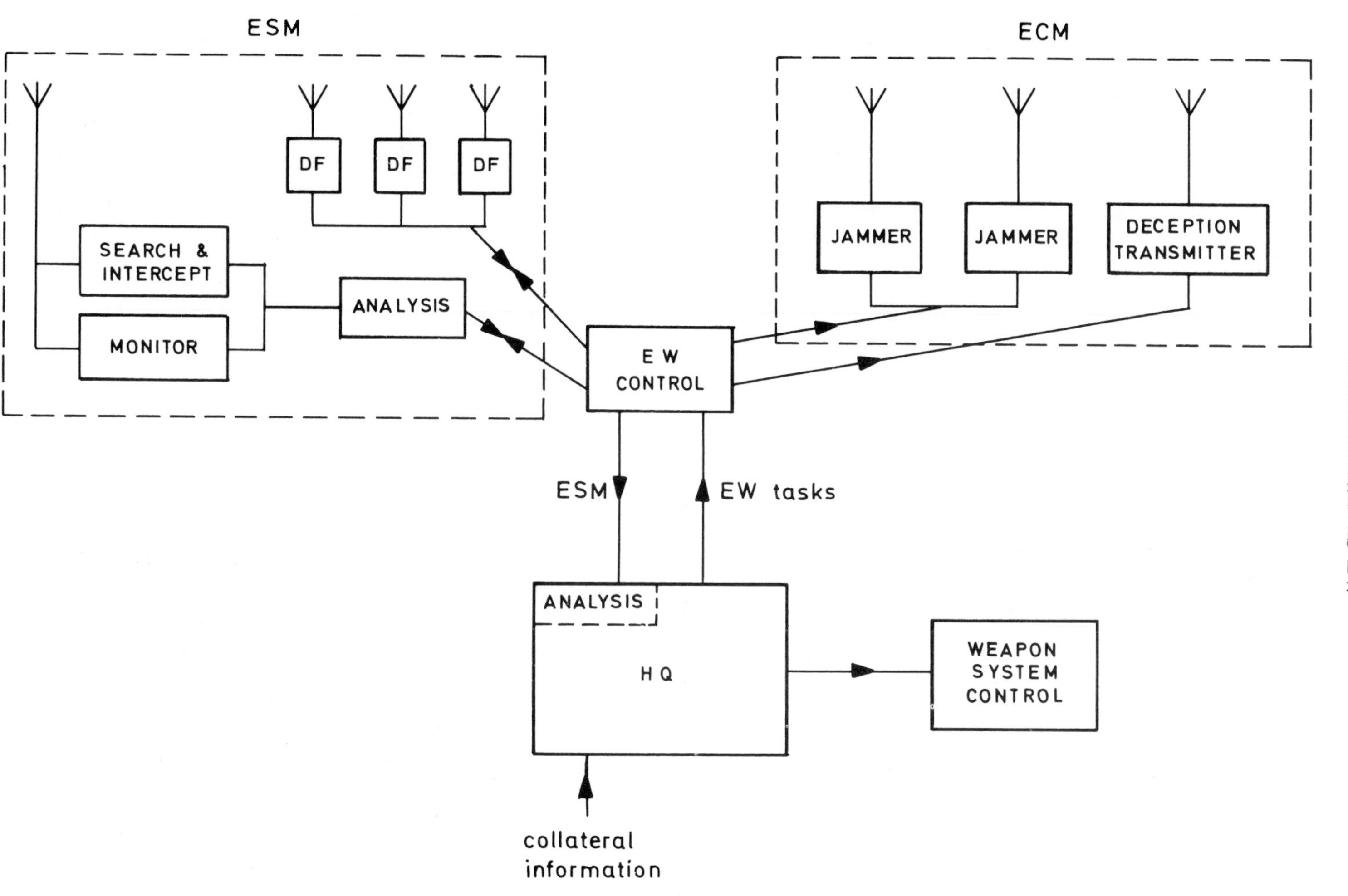

Fig. 5.2 EW system for battlefield communications

ESM is essentially the passive activity of listening in to an enemy's transmissions, whereas ECM is the active measure of attacking his communications. Figure 5.2 shows typical links between the components of ESM and the ECM and other means of attack, subsequently tasked. ESM combined with firepower, jamming and deception can provide a powerful means of attacking an enemy's command and control system.

ECCM covers the measures adopted to protect friendly communications from the EW efforts of an enemy, and may be either active or passive.

ELECTRONIC SUPPORT MEASURES

ESM, with its key components of Search and Intercept, Monitoring, Direction Finding (DF) and Analysis has the two-fold aim of acquiring operational or combat intelligence, and steerage information for offensive EW. Communications Intelligence or COMINT produces intelligence of immediate tactical value such as orders of battle, the location and identity of enemy HQs and units, movement information, indicators of enemy intentions and information on enemy electronic and cryptographic systems. It gives a commander a long range, 24 hour, surveillance capability. ESM thus provides most of the data base on which EW operations depend and as such is the fundamental component of any EW system.

Search and Intercept

A search is a reconnaissance of part of the electromagnetic spectrum to classify all transmissions occurring in it. A successful search results in the intercept of enemy transmissions. The operation can be done manually, although ADP assistance is increasingly used. It is a lengthy, painstaking procedure which must be continuous if the enemy's electronic order of battle is to be effectively evaluated but the amount of information gained can be considerable.

The search task can be made more specific by looking for particular call signs, types of modulation or other distinguishing signal or traffic characteristics. For example particular frequency bands of operation may only be used by certain nets.

The search and intercept task is formidable because there will be many enemy transmissions. They will come from many locations at different frequencies and signal strengths and will appear only intermittently. Therefore, successful intercept is heavily dependent on good equipment and intelligent siting of the ESM receiver stations. The receivers must have a good dynamic range, to handle simultaneously both weak and strong signals, over all the frequency bands of interest. It is helpful if they include digital frequency readout for simple accurate tuning, and panoramic displays, such as that shown in Fig. 5.3, to allow the operator to clearly see active frequencies, often before they are heard. In this way transmissions can be logged even when they are short, and in modern equipment this process can be automated.

Fig. 5.3 Panoramic display

There are several techniques for making a radio receiver suited for the search and intercept role. If the frequency and type of signal are completely known then optimum detection is achieved by a matched filter. This, as its name implies, is a filter designed specifically to match the wanted signal, rather than noise and interference, and it gives the best signal to noise ratio at the receiver output. To intercept more general signals it is possible to use a channelised receiver which has many narrow-band filters in parallel to cover a complete band. Such receivers can either select the largest signal present or display several signals.

Alternatively a scanning receiver, in which the tuning is swept across the band of interest, can be used and by this means the frequencies of intercepted signals can be measured accurately. However it has the obvious disadvantage that only a narrow portion of the frequency band is examined at a time. For communications, the channelised receiver approach is generally better as the complete band is monitored continuously and short transmissions are not missed. Although the frequency resolution is lower it is generally accurate enough because it is only necessary to know the frequency within the width of a single channel.

Monitoring

After transmissions have been intercepted continuous monitoring can take place to provide further information for analysis. Net activity can be logged and busy links noted. In addition direct intelligence information may be gained if clear speech is being used or if any encrypted traffic can be deciphered.

Direction Finding (DF)

After the successful search and intercept of an enemy net, direction finding provides approximate position fixes of its transmitters. The basic principle of radio DF is simple and uses the technique of triangulation to find position on a map.

Several DF receivers, at least three or four on a base line, take bearings on the target transmitter. A typical VHF DF antenna is shown in Fig. 5.4.

Fig. 5.4 VHF DF antenna

The signal is received by the four dipoles and directional information is extracted by comparing the phases of the two outputs from diagonally opposite pairs. By applying these outputs to the X and Y plates of a CRO, the operator's display will indicate bearing as shown in Fig. 5.5.

The bearings are plotted on a map to form a triangle within which the enemy transmitter should be found, as shown in Fig. 5.6.

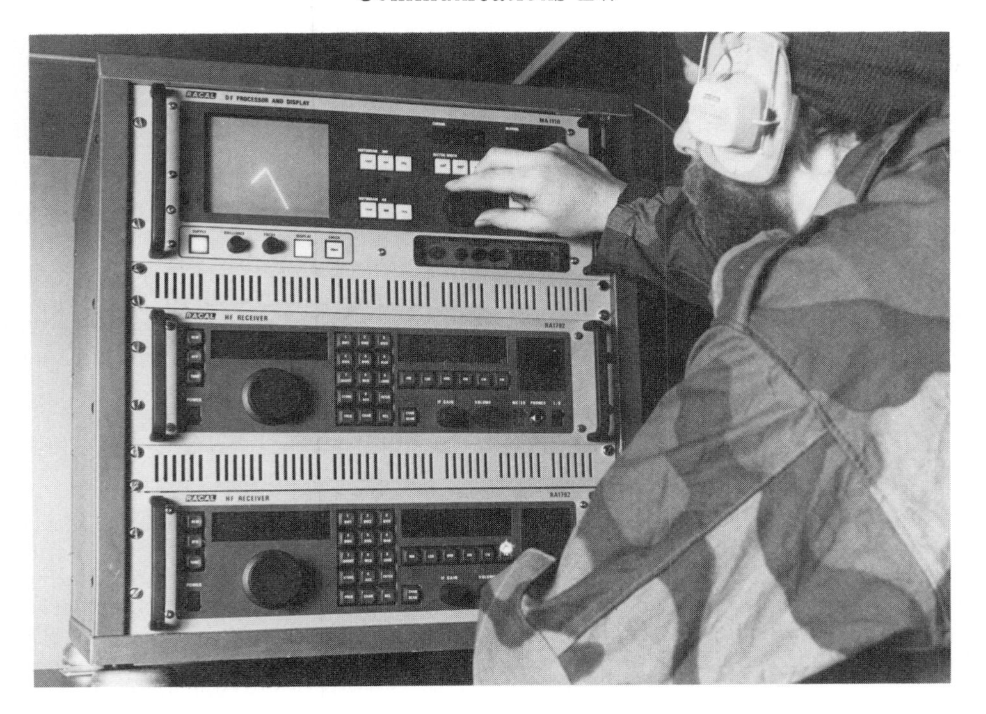

Fig. 5.5 DF bearing indication

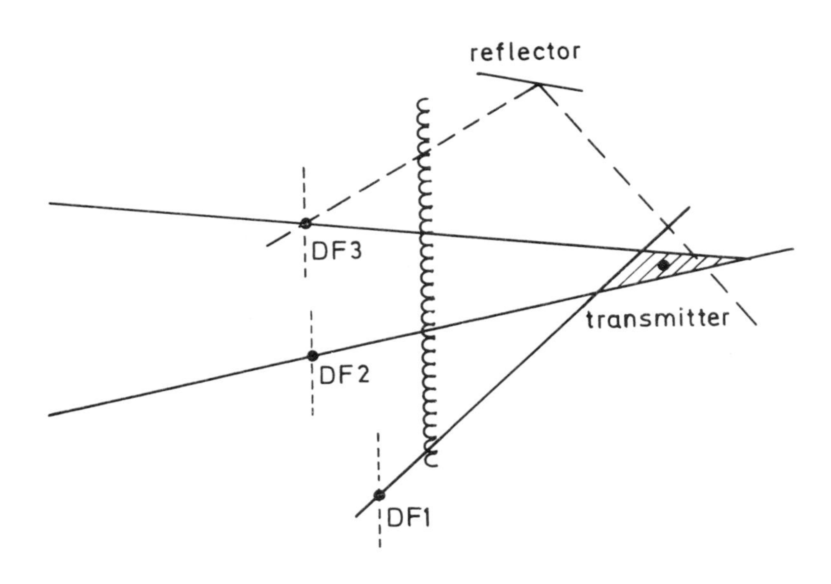

Fig. 5.6 DF triangulation

The accuracy of DF is limited by several sources of error which can be minimised but never wholly eliminated.

a. Siting. For the position of a DF antenna to be known accurately to within 10 metres an 8 figure grid reference is required.

b. Propagation variables. Ideally the path between the target transmitter and the DF station should be line of sight. Any obstructions such as hills, trees, pylons, wire fences etc may affect the transmitted signals and cause bearing errors to occur. At HF, propagation variables play a more important role than at VHF. A large reflection from a conducting surface can behave as a second transmitter of the same signal as shown in Fig. 5.6. In some cases these two signals can combine to give a single false bearing.

c. Interference. Nearby transmitters on the same frequency will interfere and produce either additional bearings or large errors.

A good DF site should be well forward and in the open. On the other hand for tactical reasons it should be concealed. The inevitable compromise is to site DF stations well forward but hidden and in common practice VHF bearing accuracies of better than 2° are difficult to achieve. This means that at a distance of 30 km there is an uncertainty of 1 km in the position of the enemy transmitter. HF DF is even less accurate, because of propagation variables. From this it can be seen that DF alone is not usually sufficiently accurate for target acquisition.

Analysis

Once DF has identified the probable area within which there is a target transmitter an intelligent map study may allow its exact location to be deduced.

An analyst can use the intelligence gained to form an overall picture of the enemy's deployment. The sort of deductions that can be made include, headquarters locations, locations of specialist units, formation boundaries, obstacle crossing points and future intentions.

The product of this intelligence gathering can be displayed in a variety of formats and map displays for specific attack systems or recce. An ESM computer can be linked to other ADP systems to provide an integrated and powerful tactical intelligence system. The process of speedy analysis is central to EW so that good use is made of the intelligence gained. The rapid growth of ADP systems is undoubtedly enhancing this process. For success, all ESM assets must be linked by a good secure communications system. This can be a problem as the communications may have to span a wide area and will also be vulnerable to enemy EW action.

ELECTRONIC COUNTERMEASURES

Whilst ESM is an entirely passive activity, communications ECM is concerned with active measures taken to reduce an enemy's use of the electromagnetic spectrum. It includes both jamming and deception.

Aim of Jamming

The aim of jamming is to prevent an enemy's transmissions being effective. A successful jamming signal will degrade the performance of the target receivers. Jamming is very much a double-edged weapon as its uncontrolled use will almost certainly conflict with friendly ESM and communication activities and even some radar systems.

To be successful the jammer must gain an overwhelming power advantage at the victim's receiver. If operating in friendly territory some distance from the target, this requires a large, powerful transmitter which is itself extremely vulnerable to the enemy's EW and weapon systems. A common method of reducing this problem is to use two jammers in a so called 'jam and scram' mode. They operate as a pair but are deployed at different locations and radiate alternately. By moving frequently their positions are less likely to be discovered by the enemy.

The battle between wanted signal and unwanted jamming signal strength is the fundamental concept behind what is known as the power battle.

Jamming Range

A typical example, as shown in Fig. 5.7, emphasises the significance of range in the power battle. The ranges and powers involved are shown. Assuming that there are no ground effects the free space loss (FSL) formula of Chapter 2 can be used.

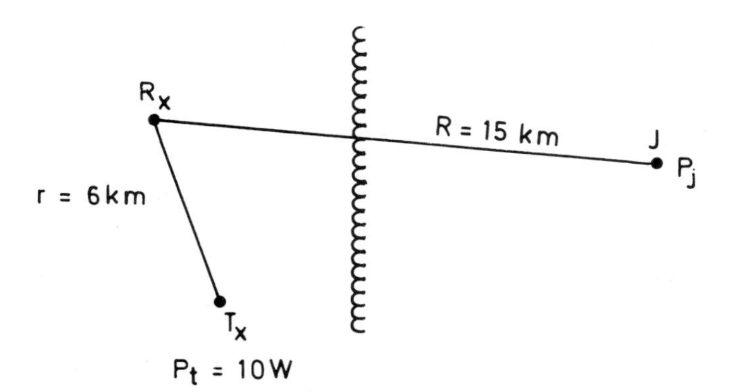

frequency = 60 MHz
λ = 5 m

Fig. 5.7 Jamming example

The wanted signal strength at the receiver is

$$\frac{P_t}{FSL} = P_t \left(\frac{\lambda}{4\pi r}\right)^2 = 0.044\mu W$$

For the jamming signal, P_j to be of the same value

$$0.044\ W = P_j \left(\frac{\lambda}{4\pi R}\right)^2$$

$$\text{so that } P_j = 62.4W$$

Thus because the jamming signal must propagate over a greater distance than the wanted signal the jamming power is high. In addition the jamming antenna is often near to the ground for tactical reasons and propagation loss rises much more rapidly than in the free space case. In this example the jamming power would be increased to 390W as ground effects attenuate the signal in proportion to $(\text{range})^4$ rather than $(\text{range})^2$.

These calculations assume that the antennas are isotropic but in practice the jammer will undoubtedly have a moderate antenna gain of at least 4, which will reduce the required jamming power requirement to about 100W.

To site a conventional high power jammer close enough to its target is often difficult, due to the constraints of the battlefield. However expendable jammers placed near the target receivers can be of relatively low power and it is possible for troops to place them before withdrawing. With current technology such devices can be highly complex, for example pre-programmed devices can be activated on set frequencies when triggered by enemy transmissions. These devices, sometimes known as 'smart' jammers can also be remotely triggered by friendly transmissions. Expendable jammers such as that shown in Fig. 5.8, can be made small and sufficiently robust to withstand gun or rocket launching and ground impact. The main weakness of remote jammers is that their battery power supply must be small and hence has a limited life. As an alternative, a jammer can be mounted on an airborne platform. This provides a good propagation path to the target receiver and the necessary transmitter power can be supplied via an umbilical cord from the ground.

Forms of Jamming

The three most common types of communications jamming are spot, barrage and swept jamming.

Spot or continuous wave (CW) jamming of a specific channel or frequency is the most common form because it causes minimum interference to friendly signals. In addition it permits optimum use of the available jamming power because the power is concentrated into a narrow bandwidth.

Barrage or simultaneous jamming of a wide band of the frequency spectrum affects a number of frequencies or channels. For a given power output it is less efficient than spot jamming as it spreads the available power over a wider bandwidth.

(Reprinted by permission from INTERNATIONAL DEFENSE REVIEW No. 4/1977. Copyright by Interavia S.A. Geneva, Switzerland)

Fig. 5.8 Expendable jammer

In swept jamming a signal is rapidly swept up and down a portion of the frequency spectrum. At any one instant only one frequency is being attacked but the effect upon receivers tuned to frequencies in the swept band may appear continuous.

Jamming can interfere with friendly communications if not carefully controlled. Indeed, as the transmissions are of high power any spurious radiation, often at frequencies other than the target frequency, can cause significant EMC problems.

It is important to know whether jamming is effective. This is achieved by continuing to monitor target nets whilst they are being jammed, by providing the jammers with a look-through capability. The jamming transmitter is switched off momentarily while the associated monitoring receiver examines the frequency band. The receiver checks to see whether communications are continuing at the same frequency or whether there are new frequencies to attack. The transmitter can then be tasked to attack appropriate enemy transmissions. Frequency changes are not always effective in evading this form of jammer. Modern jammers incorporate automatic multi-frequency jamming with a look through capability. An example is Bromure which is shown in Fig. 5.9.

Fig. 5.9 Bromure jammer

Jamming Effects

The effects of jamming differ for various types of modulation. Thus the form of the jamming signal should be tailored to suit the modulation under attack.

Frequency modulation. In FM the information is carried by the frequency variations about a centre frequency. If a CW jamming signal above a certain threshold level is received, then the jamming signal becomes the centre frequency and makes the original frequency variations invalid. The receiver is then 'captured' by the jammer. This is called 'silent jamming' as there is very little output from the target receiver. For an output to be produced by the target receiver, frequency modulation of the jamming signal is required. Typical methods are sequential tones, music, pre-recorded voice or data traffic.

Amplitude modulation. In AM the information is contained in the amplitude variations of the carrier. Thus unmodulated CW is not an effective jamming signal. A successful jamming signal should, like its target, be amplitude modulated. AM has the advantage over FM that in the presence of an appropriate jamming signal there is no threshold capture effect. Consequently AM nets can absorb considerable jamming, with a gradual degradation of signal quality, before the total breakdown of communications.

Digital modulation. If a signal is digitised then the bandwidth used is increased. Shannon's formula, in Chapter 2, showed the theoretical trade-off between bandwidth and noise and an unsophisticated jamming signal can be treated in the same way as noise. Hence, in general, as digital modulation schemes use large bandwidths they can tolerate high jamming levels before significant corruption of the received bit stream occurs.

Deception

Deception is the other active and disruptive action of EW. Whereas jamming is aimed at preventing the enemy's radio transmissions being effective, deception is used to mislead and confuse. There are two ways in which this can be done; by intruding on an enemy net or by passing dummy traffic on friendly nets.

Imitative deception. The enemy can try to deceive by entering nets and simulating their traffic. Such imitative deception can normally be detected through minor irregularities in procedure, repeated requests for acknowledgment or the inability to authenticate correctly. The intrusion that is most difficult to detect is one that is achieved by transmitting pre-recorded traffic. This is the method, which if done intelligently, has the greatest potential to create confusion.

Intrusion is more likely to succeed if communications are difficult, as the enemy can arrange to have a false message partly obliterated by interference, thus concealing any lack of knowledge of authentication or call signals. Another tactic that may be used is calling a unit with the hope of taking DF bearings on the reply. Radio operators in front line units should be especially alert for this practice.

Dummy traffic. False information or dummy traffic can be transmitted on friendly nets to deceive an enemy. For example, a dummy net using authentic procedure and passing realistic traffic can be used to hide the fact that the unit being initiated is moving under radio silence.

Prior to an operation there is often a sharp increase in the use of radio. To prevent the enemy gaining intelligence from the volume and direction of this traffic it is possible to send dummy messages to create an initial high level of radio use. As the amount of real traffic increases dummy traffic will decrease thus maintaining a steady level.

Deception control. Deception, like jamming, can be a double edged weapon and be just as disruptive against friendly forces, unless it is carefully controlled. All radio deception must be planned to fit into an overall tactical deception plan and, for this reason, it should not be attempted by individual units without reference to higher authority. The enemy's ability to achieve successful deception can be greatly reduced by alert staff users and operators operating well disciplined radio nets.

ELECTRONIC COUNTER COUNTERMEASURES

ECCM are the protective measures, applied by operators, planners and designers of communications systems to reduce the effectiveness of enemy EW activity. The aims are to prevent the enemy:

a. intercepting friendly transmissions,

b. locating friendly units and HQ's by DF,

c. gathering intelligence from intercepted transmissions,

d. disrupting friendly communications by jamming,

e. causing confusion by deception techniques,
and to mislead him over the effectiveness of his EW.

The key to effective ECCM is the prevention of enemy interception of our signals because all other activities rely on this initial intercept. Thus the primary aim is a Low Probability of Intercept (LPI) of all transmitted signals. Clearly, the most effective method is not to transmit at all! However, this is seldom practical for anything but short periods since any fluid tactical situation needs command and control.

There are two quite separate approaches to this dilemma. The first is the tactical approach where LPI is sought by a disciplined and economic use of radio which is achieved by skilful operating based upon sound training. Good tactical use of radio and the awareness by operators of the dangers imposed by enemy EW are, of course, vitally important. Much can be achieved by good tactics and procedures alone. The second approach is the technical one in which LPI is achieved by special design of radio and antenna systems.

Tactical

At the heart of any good communications plan is a sound Emission Control (EMCON) policy. This policy is formulated by a commander to suit the current tactical situation. It controls, how and when radio silence is to be imposed and lifted, the maximum power levels and antenna heights and antenna siting.

However good a formation's EMCON policy is, any radio transmission can be a source of intelligence to an enemy. It is good practice to assume that any transmission could be intercepted and the aim must be to make the enemy's task in this as difficult as possible. Hence the following tactical steps should be taken against the threat of enemy ESM.

a. Transmitting antennas should be sited carefully to avoid line of sight paths to likely enemy intercept locations and maximum use should be made of terrain screening.

b. Rebroadcast stations should be used economically especially as these are far too often placed on the best hill top sites.

c. Transmissions should be made only when absolutely necessary and for the minimum time. Even a short test transmission can be intercepted to reveal the frequency and possibly location.

d. Good net discipline should be maintained and only standard procedures and codes used. Identifiable mannerisms and deducible call signs should be avoided.

e. Frequencies should be changed frequently and at irregular intervals and can be coupled with code operator and call sign changes.

f. Radio equipment should be moved as often as possible and transmission characteristics changed if the availability of equipment allows.

g. The maximum use should be made of such alternative means of communication as line, civil telephone, dispatch rider, liaison officer, resupply vehicles and visual signals.

With the increased use of on-line cryptographic equipment there is little chance that an enemy will be capable of exploiting the content of much of the formation level tactical radio traffic. However, good EMCON is still required as intercept of secure nets can identify important headquarters from a study of traffic density.

Additional tactical measures can be adopted to reduce the effectiveness of enemy jamming and deception. For example:

a. Alternative frequencies can be included in communications planning. Their use can force the enemy to spread his jamming effort over a wider band.

b. By continuing to operate when jammed the enemy can be made to think that jamming has failed.

c. Changing the mode of transmission or increasing the power levels (but still subject to EMCON policy) the jamming may become ineffective.

ECCM TECHNOLOGY

Modern technology is enabling many ECCM techniques to be incorporated into communications equipment. In this section several of these methods will be examined.

Encryption

As described in Chapter 4 encryption of digital traffic is simply achieved by the addition of an electronically generated 'random' sequence of pulses. The key to the sequence must be non-deducible for all practical situations, it must be easily changed and protected against capture. Such encryption is restricted to digital systems which are expensive in bandwidth and this currently limits its use to VHF bands and above.

Encryption minimises the information which can be gained from traffic which is intercepted and monitored. It also prevents useful traffic analysis on trunk links because digital pulses are transmitted continuously. Thus it is not possible to detect whether or not messages are in progress. However, these are the only ECCM advantages; the signals can still be used for DF, and can be jammed. In addition an encrypted net may be immediately identified as important due to its distinctive radio signature.

Privacy

There is a requirement for short term protection of information over insecure radio and telephone circuits. Due to the short term value of information passed over tactical radio nets, forward of battalion, the full protection afforded by digital encryption is not essential. Analog speech scrambling in time or frequency can provide the necessary low quality security. For written messages, one time pad and other simple cryptographic methods, can be adopted.

Frequency Hopping

Frequency changing to avoid jamming and intercept is a basic, well practised, ECCM tactic. Frequency hopping (FH) is a natural progression and the necessary frequency agility has become possible in recent years with the advent of fast response frequency synthesisers. Figure 5.10 demonstrates the action of a FH radio.

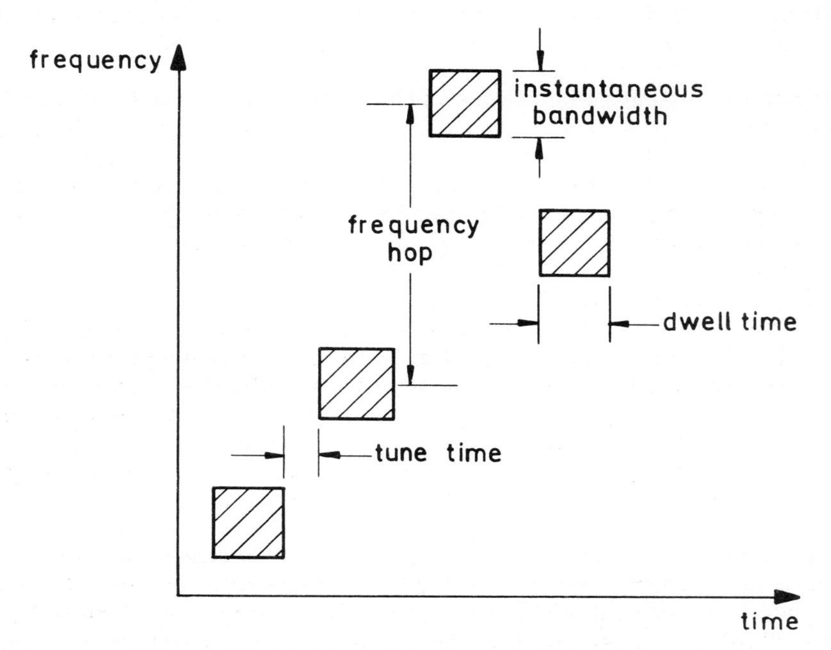

Fig. 5.10 Action of frequency hopping radio

The transmitter hops rapidly to many frequencies and the more channels used the better, since the enemy's task in searching for the signal is made more difficult. If he is forced into jamming all of them the power requires rises proportionally to the number of channels, and the advantage obtained is called the processing gain. The sequence in which the frequencies hop must appear to be random to the enemy so that the changes cannot be predicted.

In Fig. 5.10 it can be seen that not all time is spent on useful signal transmission as a tune time is required for changing the frequency. Ideally this should only be a small fraction of the time spent transmitting, which is known as the dwell time. VHF radio nets hopping at hundreds of hops per second are now in service.

The basic block diagram of a hopping system is shown in Fig. 5.11 overleaf. The action of such a system is similar to the conventional net radio described in Chapter 3, except that the oscillators or frequency synthesisers, in the transmitter and the receiver are controlled by sequence generators. The sequence generators change the frequency in a random manner determined by an electronically generated pseudo random code. At the receiver, synchronisation between the received waveform and the internally generated code is essential and is achieved by a sync detector circuit. For net radio such synchronisation must be achieved rapidly and automatically. This applies to both current and new, or late-entry users even when the received signal is weak or jamming is present. Synchronisation and information data are interleaved to ensure that this process is continuous.

There are, as yet, no defined standards for the hopping rate but rates of less than 10 hops/sec are referred to as slow, 100 hops/sec as medium and rates in excess of 1000 hops/sec as fast. Because of the severe propagation variables at HF only slow hop rates can be used. At VHF and UHF most emphasis is being placed on medium speed hoppers.

At VHF slow hopping systems offer no significant advantages as it is possible to follow and jam. Even spot frequency jammers can be very effective because at these rates each hop is about a syllable in length and may be completely lost. For fast hoppers synchronisation is complex and is inefficient because the time required to change frequency becomes significant. Consequently, present development is concentrated on medium hoppers since these are extremely difficult to follow and jam.

Spectral pollution refers to the large number of unwanted frequencies within the radio spectrum and is an important problem with frequency hopping radios. Many frequencies are used directly and the problem is further aggravated by the harmonics produced by the abrupt changes of frequency. This effect is minimised by the smooth reduction or increase of power during frequency changes. The frequency hopping net must be compatible with nets using conventional single frequency radio. This is achieved by barring some frequencies.

It is possible for all nets to hop over the entire VHF band and this could cause internet interference if different nets hopped to the same frequency. Such a wide band approach makes intercept difficult but frequency management has to be carefully controlled. There are two methods of frequency planning. One

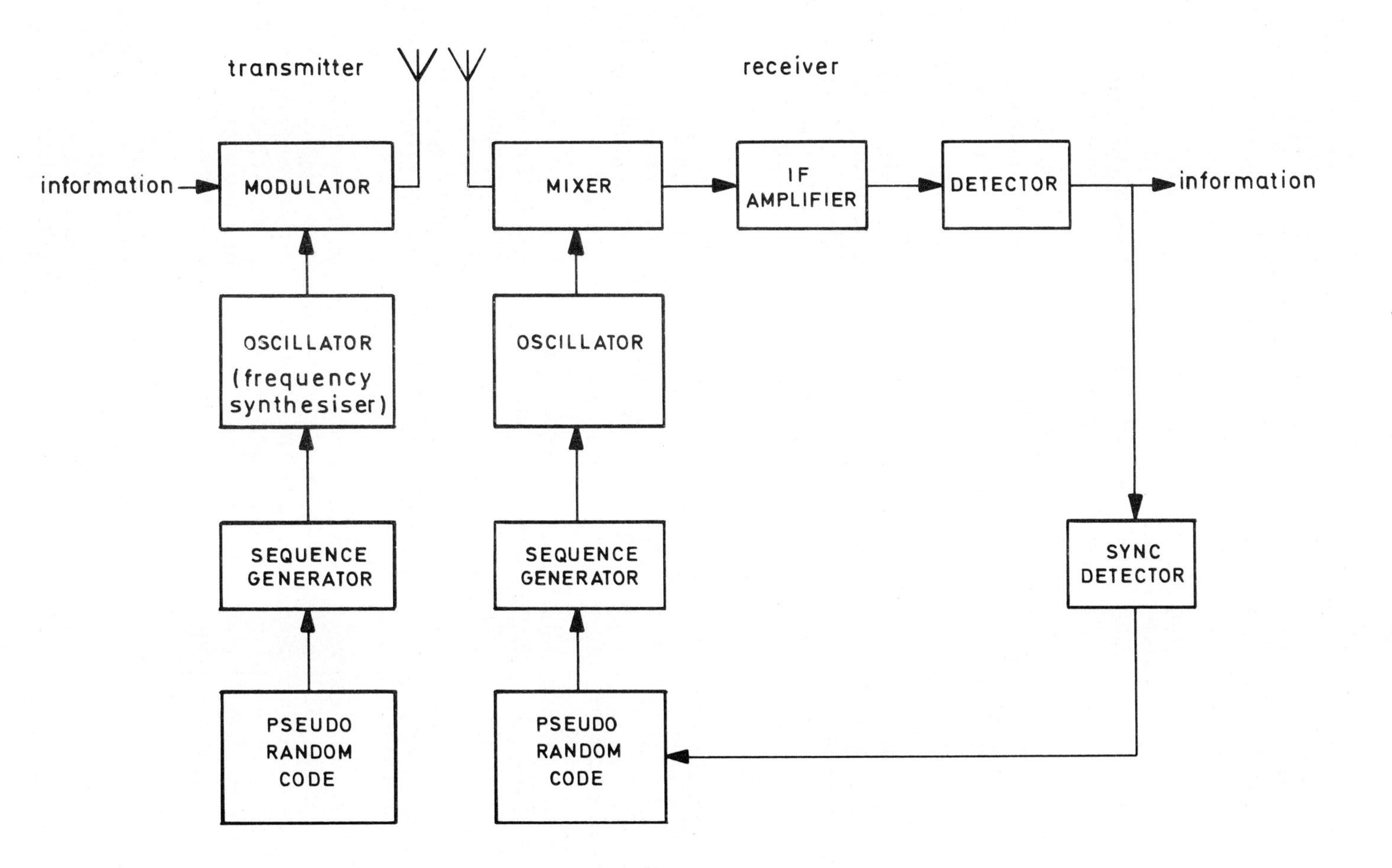

Fig. 5.11 Frequency hopping system

approach is to divide the available frequencies into bands a few megahertz wide used only by a few nets. The mutual interference problem is reduced by this technique, but, because each net is restricted to a band it will have a distinct signature. This method is called non-orthogonal hopping and within an allocated hop band momentary interference between nets is possible. Timing control is achieved by one station being designated as the master and other stations as slaves. The second method is orthogonal hopping in which interference between hopping nets is eliminated by ensuring that any one frequency is used by only one net at any time. In this method the whole band may be used and thus the jammer is forced to attack all frequencies. The timing for such a hopping system is clearly critical and very accurate clocks are used for synchronisation.

Hopping radios have the advantage of being able to work through considerable interference but there are some operational limitations. Siting is difficult due to poor EMC performance compared with conventional radio. Congestion of the frequency spectrum, in an already crowded environment, poses planning difficulties. An inherently high error rate makes hopping systems unsuitable for data unless considerable error protection techniques are used. As a general rule, a single hopping net should not be fielded since it produces a distinctive signature and is an attractive target. On the other hand, the use of many nets reduces the risk of intercept of any one net, as the overall radio picture is confused.

Direct Sequence Spread Spectrum

Normal and FH VHF net radios use FM which is straightforward to intercept because the signal stands-out above the normal noise background. Direct Sequence Spread Spectrum (DSSS) spreads the signal energy over a very wide band before modulating to a radio frequency and typically a 3 kHz voice bandwidth signal is spread over a 2 MHz bandwidth. The process gain is the ratio of these bandwidths $(2 \times 10^6 / 3 \times 10^3)$ and a jammer must increase power by this amount to be successful. As the energy density over the range of frequencies is significantly reduced it is difficult to distinguish the signal from background noise and hence difficult to intercept.

Figure 5.12 shows a diagram of a DSSS system. The pseudo-random code switches the signal by 180° (π-phase switching). If the code bit rate is 1 Mbit/sec the resulting radio frequency signal occupies a bandwidth of 2 MHz. If the code is pseudo-random the radio transmissions appears as noise. In the receiver an identical and synchronous code despreads the signal by effectively cancelling out the phase reversals. The result is an intermediate frequency version of the original signal. The main advantage of this technique is that other, interfering and jamming signals contained within the radio frequency bandwidth, suffer from the high-speed reversals in the receiver. Thus the interference is spread while the wanted signal is despread.

In combat net radio the direct sequence spreading technique appears to offer a low probability of intercept (LPI) and the ability to reject jamming. A major problem is that the receiver can be jammed by a nearby spread-spectrum transmitter if its power exceeds the wanted signal plus the process gain, even though this transmitter is using a different code. This is called the near/far effect and limits

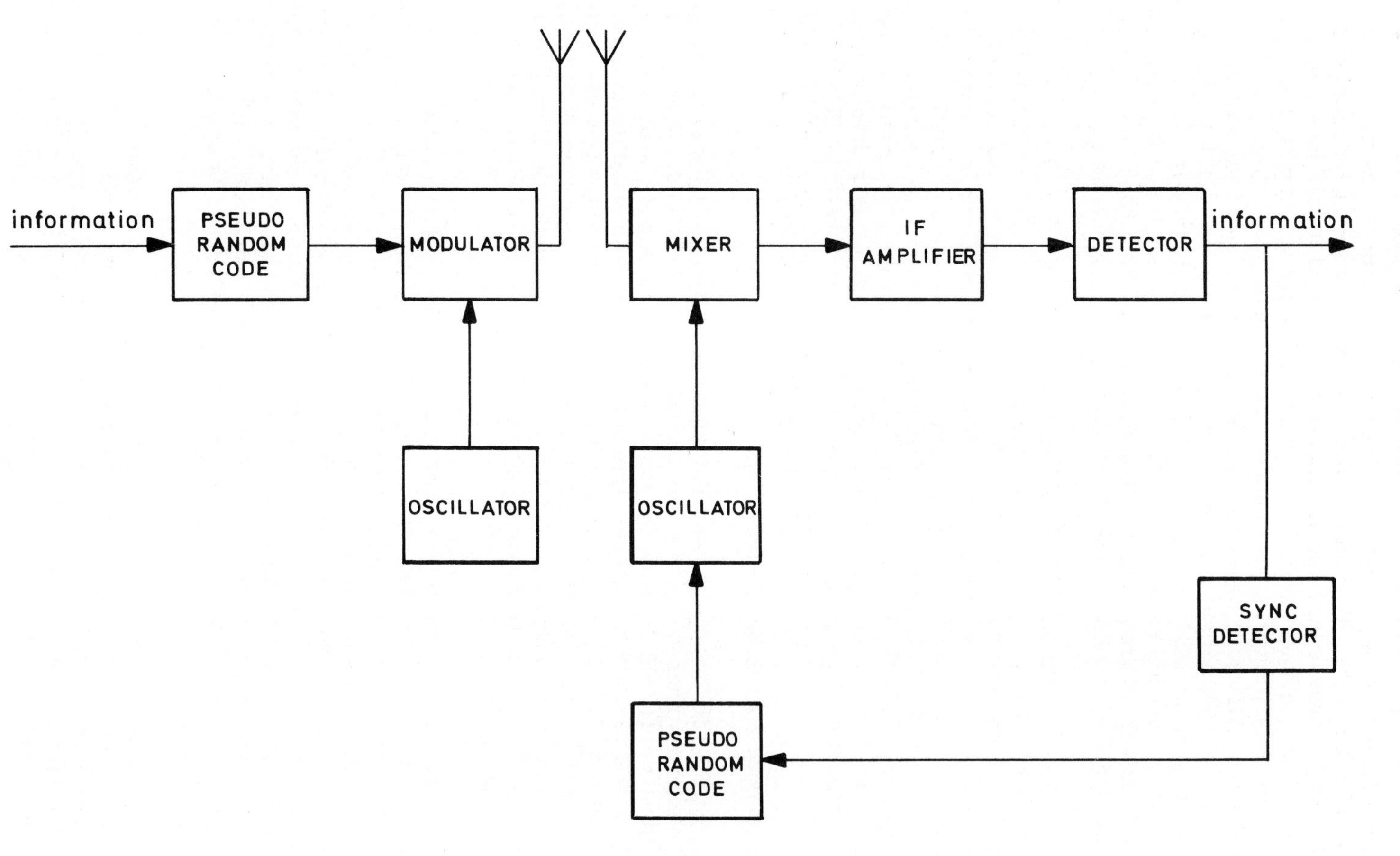

Fig. 5.12 Direct sequence spread spectrum system

the use of this type of modulation to a link rather than a net. However a unique advantage of DSSS is that it could be used in covert operations, or during periods of radio silence, because its signal appears as low power noise.

Null Steering and Adaptive Antennas

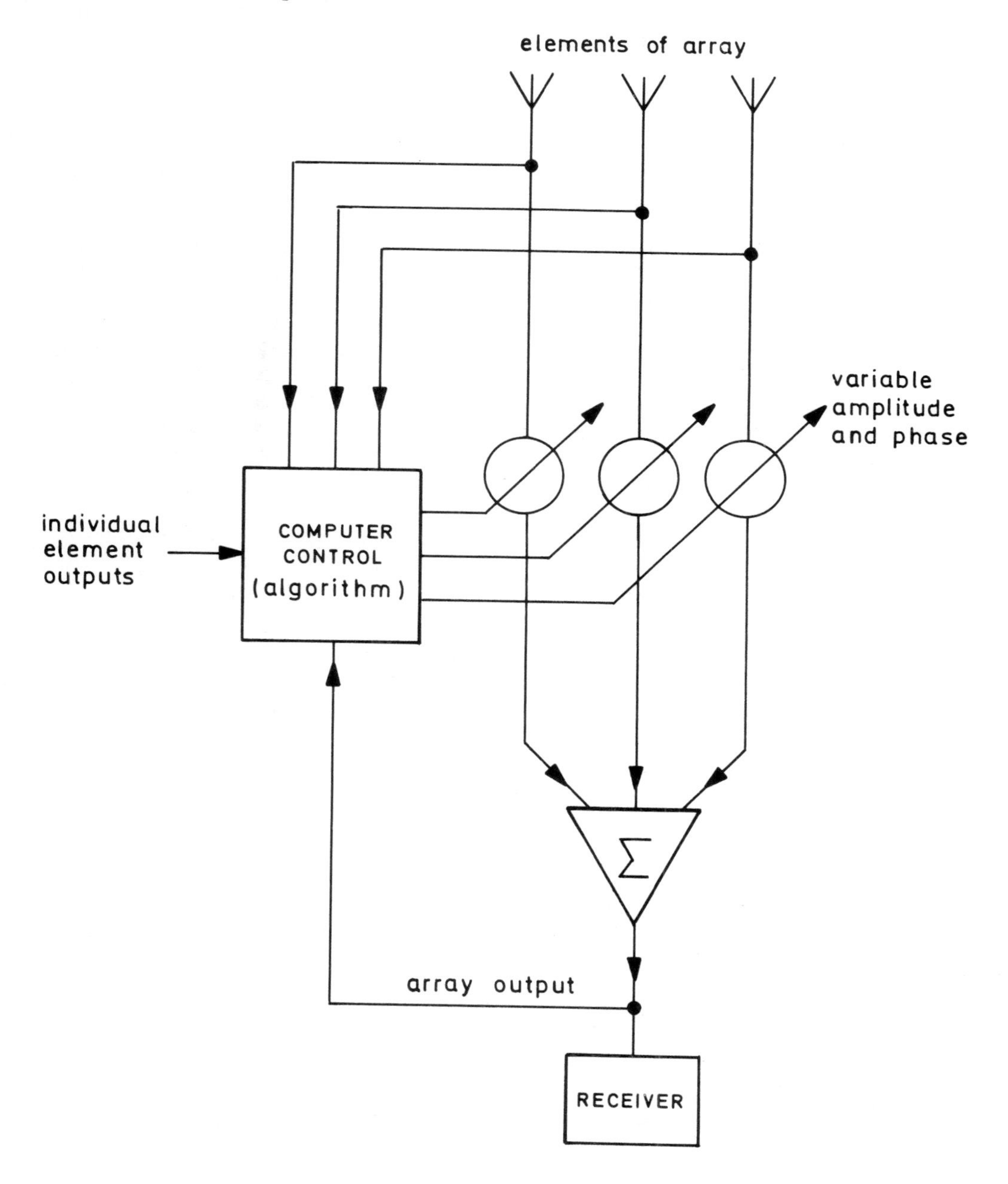

Fig. 5.13 Adaptive antenna system

A problem with the simple single whip antennas used in tactical net radio is that their radiation patterns are omnidirectional and such indiscriminate radiation clearly increases the probability of intercept. The antennas are used for both transmission and reception and as they have identical patterns in each role they are also vulnerable to jamming.

Although antennas can be designed to have nulls in their patterns, changing jamming situations and other forms of interference, make it desirable for these to vary automatically. In the 1960's the Sidelobe Canceller was developed to reduce the effect of jamming on radars and it has subsequently been used in some communication cases, such as high gain satellite ground stations. A low gain auxiliary antenna is used to receive an independent interference signal which is then subtracted from the main antenna output. This system relies on the interference being the largest signal received by the auxiliary antenna.

The more complicated fully adaptive antenna uses an array of usually identical elements as shown in Fig. 5.13. Measurements of outputs are made and from these an algorithm (computer program) attempts to find how the individual element outputs should be combined in order to minimise interference. If N elements are used, then the array can handle either N-1 interferers at one frequency or fewer broad-band interfering sources. However, the cost of the system can become large because, in addition to the measurement and control equipment, every element requires circuits to vary the amplitude and phase rapidly and accurately.

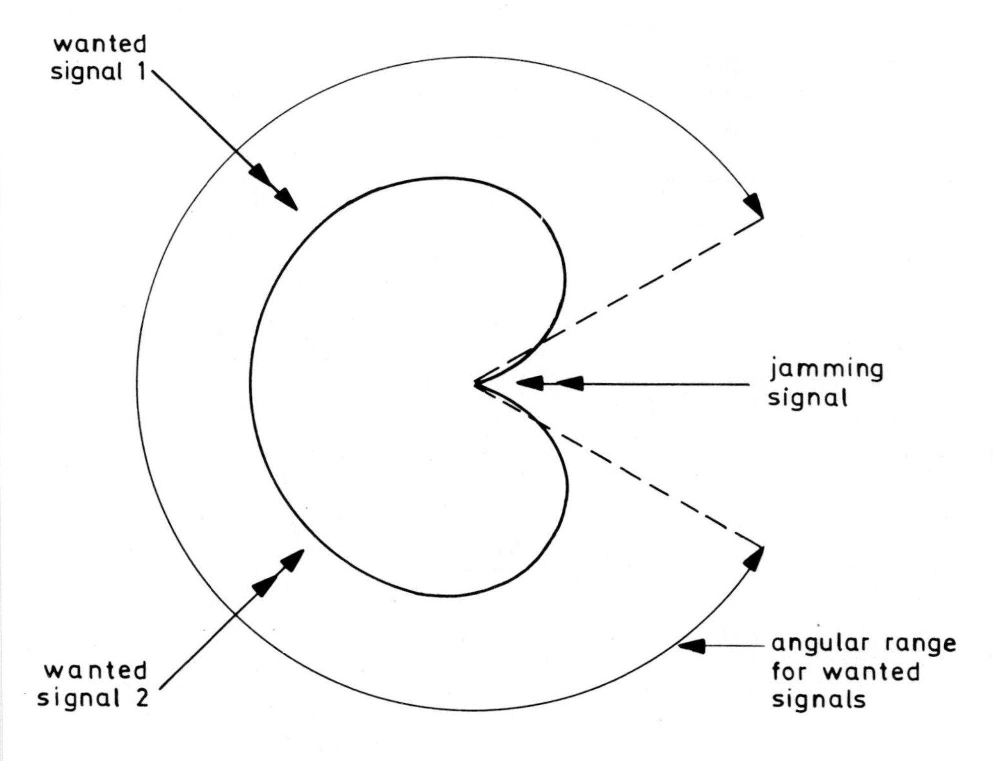

Fig. 5.14 Nulling of jamming signal

Another practical problem can be the physical size of a multi-element adaptive antenna as for useful pattern nulls, element spacings around half a wavelength are required. Thus for tactical communication at VHF small two or three element arrays are probably the largest feasible size. In these cases the single null of a cardiod like pattern shown in Fig. 5.14 can be directed towards a jammer without significant degradation in wanted signals. The steering can be automatic or manual, this latter possibility being much cheaper to implement, yet giving effective nulling. A practical system can reduce the effect of a jamming signal by at least 30 dB. A further use of an antenna pattern null is to point it in the general direction of the enemy to reduce the probability of intercept.

The combination of null steering antennas and a friendly jammer can be a powerful EW technique. Figure 5.15 shows two radios, which are part of the same net, and can communicate because their antennas are nulling the effects of the friendly jammer. However the enemy has to 'look' towards the jammer to intercept the net and as the jamming signal is likely to be the largest, interception is difficult. This technique is often referred to as 'smoke screen' jamming.

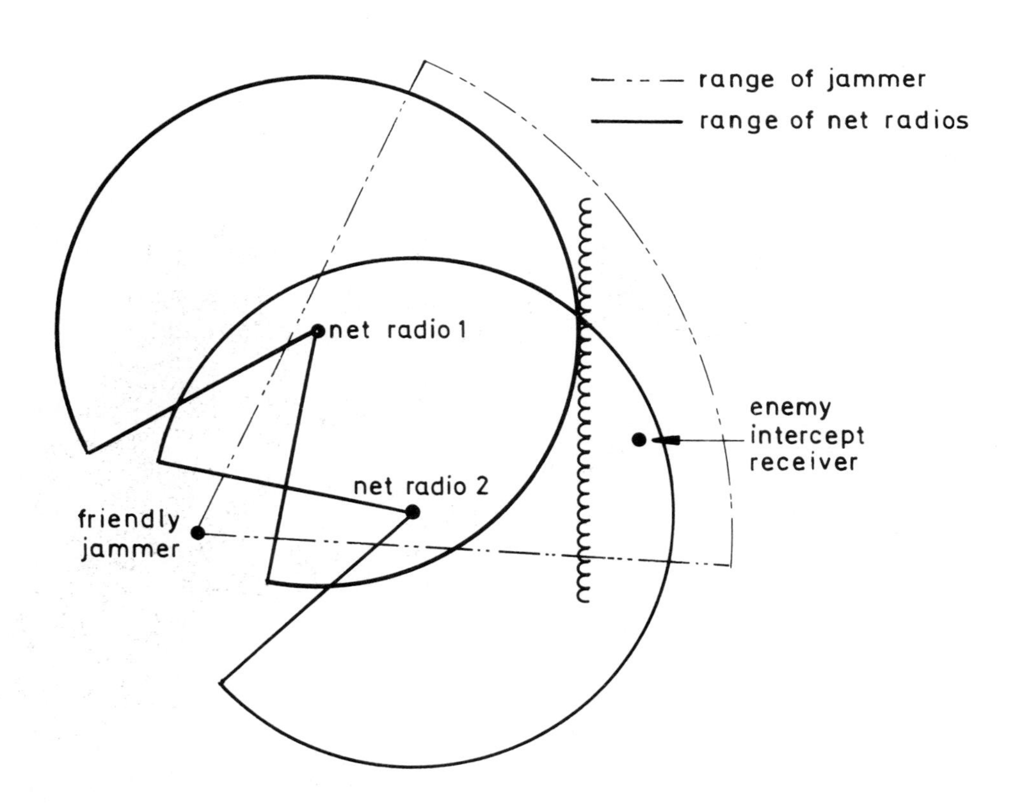

Fig. 5.15 Null steering + jammer EW technique

Burst Transmission

Communicators have a certain bandwidth available for a period of time and must decide how to best use this potential capacity. The most common method is to use a fairly narrow band for a long time. Alternatively a much wider band can be used for a short burst. The short transmission time makes detection and DF difficult, particularly if the frequencies used are changed frequently.

To transmit by bursts it is necessary to assemble messages as data and it is this step that leads to its ECCM advantage. Data is a much more compact way of passing information than speech, because in assembling data, the user is forced to work out his message precisely and so wastes little of the communications channel 's potential capacity. Data is ideal for clearly formatted material such as grid references, lists and contact reports. Although speech is important in many circumstances, the use of burst data where appropriate would overcome much of the congestion in current net communications.

Figure 5.16 shows a Message Entry and Read Out Device (MEROD). The messages are entered via a keyboard and stored in an electronic memory so that checking and modification is possible. The entry device automatically formats the

Fig. 5.16 MEROD burst transmission

message and adds bits to allow error detection to take place at the receiver. If necessary the burst can be repeated until it is received correctly. The data stream can be interfaced to the audio socket of a conventional radio and the transmitted and received messages are read on the display. The data is transmitted over the most appropriate bandwidth. At HF the narrow band option enables interference free channels to be found where there would be difficulty in finding suitable speech width channels. The broadband, short time alternative makes intercept difficult and if bursts are sufficiently small they do not significantly degrade continuous speech using part of the same band.

Single Frequency Rebroadcast

The conventional method of rebroadcasting to extend the range of VHF nets, is to use a second non-interfering frequency as was described in Chapter 3. A revolutionary design has now made automatic single-frequency rebroadcast possible although it is presently limited to low power, FM, which effectively limits its tactical use to forward nets. However it could play a significant role in improving ECCM. Figure 5.17 shows how such a rebroadcast station can complicate the enemy's problem in finding a headquarters by radio DF.

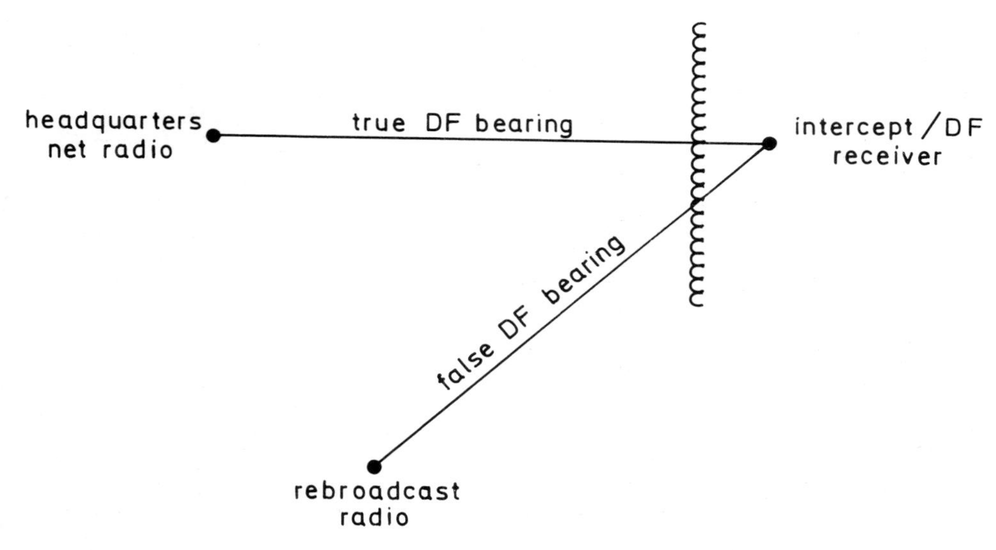

Fig. 5.17 Single frequency rebroadcast

The use of this type of rebroadcast radio extends the coverage of a tactical net without using additional frequencies. However, a serious EW disadvantage is that a jammer can also effectively extend its range if the rebroadcast radio is electrically captured. Single frequency rebroadcast is one of a number of adaptive cancellation methods which can be used to improve net radio. The techniques and their uses will be covered in more detail in Chapter 6.

Summary

Table 5.1 gives, in condensed form, the strengths and weaknesses of these technical methods for improving ECCM performance.

TABLE 5.1 Technical ECCM Techniques

Technique	Advantages	Disadvantages
Encryption	Total security of information.	Increase in bandwidth. Has a distinct signature.
Privacy	Small and cheap equipment. Little increase in bandwidth.	Suitable for short term security only. Has a distinct signature. Some loss of quality in speech.
Frequency Hopping	Difficult to intercept, DF and jam.	Uses wide bandwidth. EMC difficulties, particularly with multi set vehicles. Single net has distinct signature.
Direct Sequence Spread Spectrum	Hides signal in noise, difficult to intercept. Resistant to jamming.	Very wide bandwidth. Near/far effect limits use.
Null Steering Antennas	Reduces susceptibility to intercept and jamming. Can be used in conjunction with smoke screen jammer giving powerful ECCM advantage.	Complex and expensive multi-element antenna control.
Single Frequency Rebroadcast	Extends area coverage with no extra frequencies needed. Effective deception tool, countermeasure to enemy DF.	Extends, if electrically captured enemy jamming range. Limited to low power only.
Burst Transmission	Reduces time on air, difficult to intercept and DF.	Not suitable for speech.

DISCUSSION

It can be seen that EW adds another dimension to the battlefield. It shares the same goals as the physical battle and it is, in most respects, governed by the same military principles. Domination of the electromagnetic spectrum is, with the increasing dependence of commanders on C^3 systems, proving every bit as important as domination of ground and forms an essential part of any commander's plan. Much can be achieved by good communications tactics alone. It is clear that a pre-requisite of survival against a technically advanced enemy is an appreciation by all users, of the potential EW threat. If the tactical consequences of such a threat are remembered by all users before the transmit-button is pressed, then much will have been achieved.

A great deal can be achieved by making use of recent advances in electronic technology. Indeed, if C^3 is to continue to be a potential force multiplier then our communications systems must use every advantage offered by new developments.

It is important to appreciate that any advantage gained, whether procedural or by using new technology, is likely to be a temporary one. Furthermore, as all the techniques have significant vulnerabilities as well as advantages it is most unlikely that any one technique will dominate.

Furthermore, in designing any new range of tactical communications equipment, consideration should be given to include several techniques. This can be achieved by building in modular form with the maximum use of signal processing and microcomputers. Such systems have the significant advantage of flexibility and ECCM techniques can be selected to suit a particular set of circumstances. This approach also forces the enemy into developing a similarly wide range of countermeasures.

SELF TEST QUESTIONS

QUESTION 1 What characteristics should a good ESM receiver have?

Answer ....

....

QUESTION 2 What COMINT can be gained from ESM?

Answer ....

....

QUESTION 3 Discuss the siting of DF stations.

Answer ....

....

QUESTION 4 What are the advantages and disadvantages of remotely delivered expendable jammers?

Answer a. Advantages ....

....

....

b. Disadvantages ....

....

....

QUESTION 5 What ECCM tactics can:

a. Minimise the possibility of intercept?

Answer ....

....

b. Reduce the effects of a jamming signal?

Answer ....

....

c. Deceive an enemy jammer?

Answer ....

....

d. 'See-through' deception techniques?

Answer ....

....

QUESTION 6 Discuss the difficulties associated with introducing frequency hopping for net radio communications.

Answer ....

....

....

QUESTION 7 What further techniques are worthwhile investigating to improve the ECCM characteristics of net radio?

Answer .. ..

....

ANSWERS ON PAGE 137

6.
Other Communications Systems

INTRODUCTION

In the earlier chapters the principal tactical communication systems of net and trunk radio, have been described. However there is also a military requirement for strategic communications, both in local areas and worldwide. The aim of this chapter is to complete the description of military communications by considering alternative communications techniques and describing how they can be used to satisfy the complete range of command and control requirements.

The section on techniques contains descriptions both of the means for providing complete links and the methods of improving the performance of conventional radio systems. Several of these methods use comparatively new technology and it is likely that their importance will grow in the future. This is followed by a section in which the requirements of strategic systems are examined and the communications methods being employed, now and in the near future, are noted. In addition tactical systems are re-examined to show how they can be modified and enhanced by the use of modern techniques.

SECTION I - TECHNIQUES

SATELLITE COMMUNICATIONS

A communications satellite is essentially a microwave radio relay station in the sky and can be used for over-the-horizon communications. The satellite receives signals in one frequency band and retransmits in another, typical military frequencies are 8.5 GHz for the uplink and 6.5 GHz for the downlink. The receiver/transmitter combination of the satellite is usually known as a transponder. Figure 6.1 shows an artist's impression of the SKYNET IV satellite which is intended to be launched in the mid eighties. The satellite derives its electrical power from the large panels of solar cells which unfold to the positions shown, once in orbit. The satellite is oriented to point its antenna platform towards the ground and the

Fig. 6.1 SKYNET IV satellite

antennas illuminate appropriate areas of the earth. Note that in this example the dishes have multiple horns at their focii and the outputs, from these horns, can be combined to give alternative coverage patterns or 'footprints' on the ground; this is described in the later section, on SDMA.

Orbit

Most communications satellites orbit at a height above the earth of 35,860 km. These are known as geosynchronous orbits since the satellites travel around the earth in exactly the earth's rotation time. If its orbit is over the equator then a satellite appears to be stationary over one point on the earth. This geostationary orbit has many advantages in that:

a. There is no need for sophisticated tracking by the ground stations.

b. The satellite is permanently in view.

c. 42% of the earth's surface can be covered by one satellite or global coverage can be provided by three satellites.

d. There is virtually no Doppler shift (ie frequency shift due to movement) to be removed from the received signals.

The main disadvantages are that:

a. The received signals are weak because of the great distances involved.

b. The propagation delay from one earth station to another is 270 milliseconds.

c. The number of suitable positions for satellites in space is limited and already the geostationary orbits nearest to the developed nations are becoming congested.

Geostationary satellites have their orbits adjusted occasionally because of minor gravitational perturbations caused by the varying positions of the sun, moon and earth. This adjustment is achieved by releasing hydrazine gas under pressure.

Power Budget

For any communications link there is a minimum signal-to-noise ratio that must be exceeded for satisfactory communication. The sum, in dB, of gains and losses in the link is known as the power budget and it is possible to design the various components to produce the most convenient overall system.

The transmission path from the ground to a geostationary satellite is 36,000 Km and the free space loss is about 200 dB at typical satellite frequencies. In addition atmospheric absorption can add several dB's of loss depending on frequency, weather precipitation and angle of elevation; at low elevation there is a longer path through the atmosphere.

High gain antennas guide power into a narrow beam, as described in Chapter 2. Thus the effective power, in the desired direction, is increased. For high gain, an antenna must be electrically large and this requires either that it is physically large or that the frequency used is high. However at high frequencies transmission losses rise and cancel out much of the increased antenna performance. In addition the narrow beamwidths of high gain antennas make them difficult to point in the correct direction.

At the receiver, signals can be amplified to compensate for the transmission losses but it is the signal-to-noise ratio which is most important. The sources of external noise are galactic, atmospheric and man made, and their levels depend on the temperatures of their sources. For example a link from a satellite can become unusable when the ground station antenna, due to the relative position of the satellite, is forced to point at the sun. It is also important that the receiving antenna and receiver electronics introduce as little noise as possible. Such losses are minimised by connecting a low-noise receiver as close to the antenna feed horn, as possible. The effectiveness of this combination is often quoted as a ratio of gain to temperature or G/T. The noise of a system is also directly proportional to the bandwidth. Thus adequate signal-to-noise ratios are easier to achieve for low capacity, narrow band communications.

The trend in satellite technology has been for the satellites to become progressively larger thanks to more powerful launch rockets and the space shuttle. The size of solar panels has increased so that more power can be transmitted from the antenna. The antennas are also physically larger, more efficient and of higher gain. By considering the trade offs of the power budget this leads to two forms of improvement. First, larger bandwidths can be used so that speech links are possible where previously only telegraphy could be achieved. Second, less gain is required at the ground station so that ground stations can have smaller antennas which, due to their broader beamwidths, are easier to point. Figure 6.2 shows a prototype model of a military manpack satellite ground station.

Multiplexing and Multiple Access

A satellite can be used to provide a point-to-point link between two earth stations. As in trunk radio relay, if this link has a wide bandwidth it can be shared by many channels by the use of frequency or time division multiplexing as described in Chapter 4. However an additional feature of a satellite is that it has a wide area of coverage and so can be shared by geographically dispersed earth stations. This form of sharing is accomplished by multiple access techniques which dynamically assign channels according to user demand.

Frequency-Division Multiple Access (FDMA). In this technique different carrier frequencies are used by each ground transmitting station. This enables many stations to share a single wideband satellite transponder until saturation, due to excessive power, occurs. The use of the electronic circuits by many carriers results in increased intermodulation which raises the apparent noise level. This occurs because when the circuits are saturated, received frequencies modulate each other to produce false frequency components within the transponder bandwidth. To minimise this effect power levels are reduced to prevent saturation occurring.

Fig. 6.2 Satellite manpack ground station

Time Division Multiple Access (TDMA). All earth stations use the same carrier frequency but each station is allocated time slots for its use alone. This technique is generally superior to FDMA because intermodulation noise is lower and the total satellite power and bandwidth is available to each user. However accurate timing and synchronisation of the users is required and this is particularly difficult to control when there are many mobile users.

Space Division Multiple Access (SDMA). This technique requires multiple spot beams from the satellite antennas so that the ground stations are illuminated by separate beams. The satellite can switch beams on and off and switch channels between beams. Satellites with large complex antennas are required to produce these beams but even then the technique is only suitable for widely separated ground stations.

Code Division Multiple Access (CDMA). Sometimes called Spread-Spectrum Multiple Access, this technique requires each user to be identified by a code. Information is combined with a pseudo random sequence, derived from the code, and this causes the transmission to be spread out and occupy the entire bandwidth of the satellite transponder. If the receiving earth station is able to duplicate the pseudo random sequence then when synchronised with the received signal, the original information can be extracted. All other signals are based on different pseudo random sequences and appear as background noise. This technique is advantageous in military systems because it has a low probability of intercept and an anti-jamming capability.

To use the potential of a satellite fully, different sequences of terrestial multiplexing, carrier modulation and multiple access can be used. It could, for example, use a sequence such as FDM/FM/FDMA.

LINE COMMUNICATION

At least part of a telephone or telegraphy circuit will employ lines or cables. In many civil, and certain static military systems, the cable network extends over considerable distances. The two most significant electrical properties of a cable are its attenuation, measured in dB/km, and its bandwidth. In general reducing the loss or increasing the bandwidth requires a larger and heavier cable.

Static trunk systems use coaxial cables to carry groups of telephone channels over long distances. Tactical trunk systems generally use radio for long links but cables are used for the tails, which are the links from headquarters to radio relay sites. In addition similar links connect net radio to the headquarters. These lines are laid before the radio communications and the staff move to a new site. After a move lines are recovered and used again. The cables should be as light as possible, consistent with providing an acceptable bandwidth and loss because the quantity of field cables can be large and requires several vehicles for its movement and deployment.

Within a headquarters multicore cable is used because many short links are required to connect vehicles together. To simplify wiring the various channels are grouped logically to multiconnector sockets on the skins of vehicles and interconnection is straightforward.

Line has certain advantages and disadvantages compared with radio. Although it radiates, the level is much lower and harder to detect than a radio link. However lines are vulnerable to enemy sabotage and covert interception so that tails need to be patrolled and protected.

If cheap, lightweight cable is available then dispensable/disposable links are feasible. Rapidly laying such links from helicopters can provide a useful alternative to radio particularly when faced with hostile EW. Pre-wiring locations can also be carried out if battle moves can be predicted although there is the risk that such lines may be tampered with by the enemy before they are brought into use. A fundamental consideration of the greater use of line is now taking place because high bandwidth, lightweight fibre optic cables are rapidly replacing conventional copper cable in civil trunk systems.

Fibre Optics

Transmission at light frequencies is playing an increasing role in communications. Light can be guided along small diameter optical fibres (2-200 microns), and low

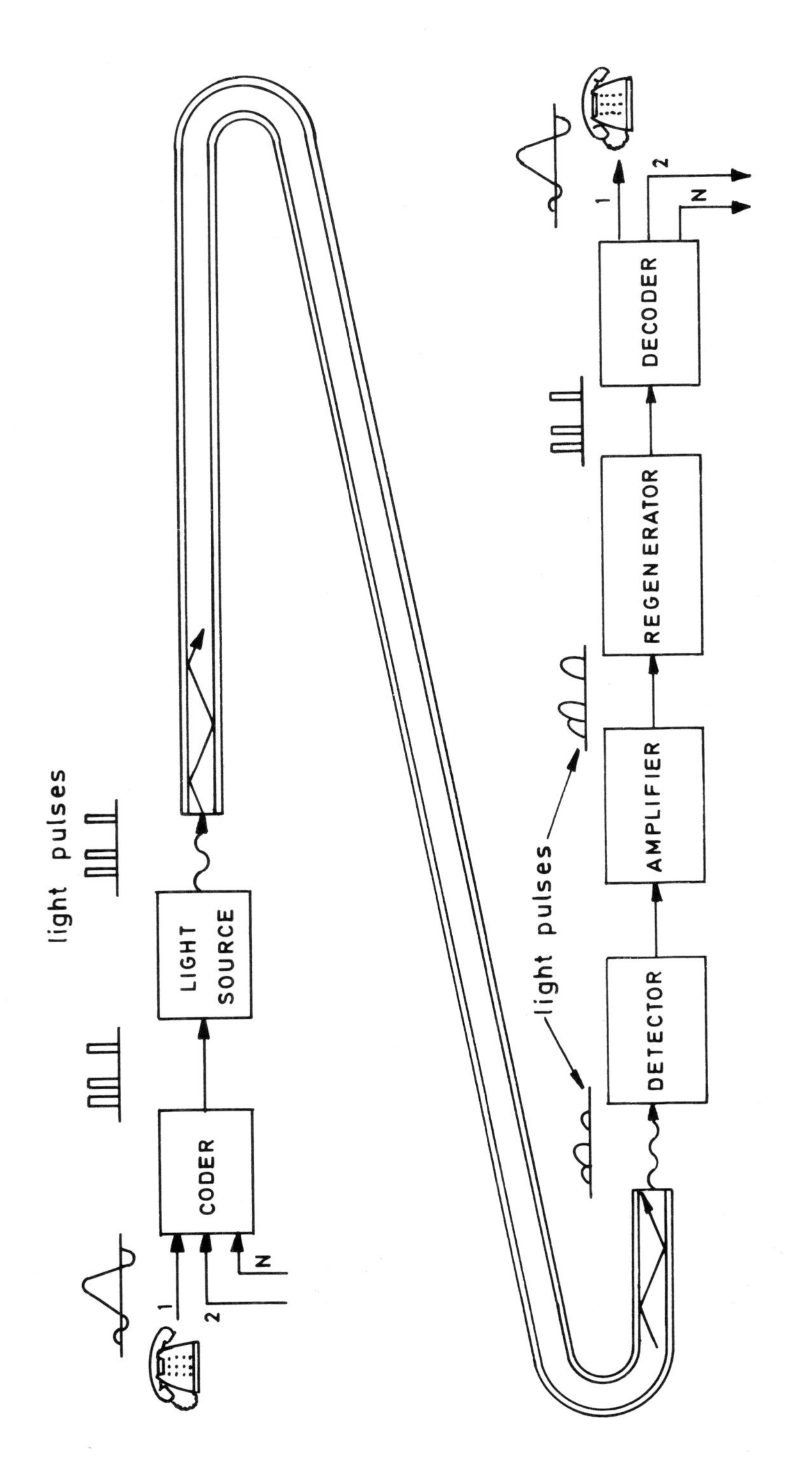

Fig. 6.3 Fibre optic system

loss and high bandwidths can be achieved. The performance and costs are improving rapidly and it is cost effective to use optical fibres for civil high capacity trunk telephone circuits.

Figure 6.3 illustrates a fibre optic system. A bit stream is produced in the same way as in any other digital communication system. A semi-conductor light source is modulated to produce a stream of light pulses which enter the fibre. Due to the characteristics of the fibre these pulses are attenuated and distorted. At the receiver a photo diode converts the pulses back to electrical signals which are then amplified and regenerated to reform the original bit stream. The key elements of the system are the fibre cable together with its connectors and splices, the light source and the photo diode detector.

Compared with earlier methods of carrying communications traffic, fibre optics offer high capacity from very lightweight cables. For military use they offer the further advantages of transmission security, as they do not radiate, and radio frequency noise immunity. Current developments are improving performance and lowering costs whilst at the same time the cost of conventional copper cable continues to grow. In the future the use of fibre optics will be widespread.

Fibre Types

Fibres can be manufactured from glass or plastic. Although plastic fibres are extremely rugged their high attenuation limits them to lengths of tens of metres only. Telecommunications quality cables are made from very pure glass and the degree to which impurities are controlled during manufacture determines the attenuation of the fibre. Light is totally internally reflected at the boundary between regions of glass which have different refractive indices. There are three types of fibre, multimode, monomode and graded index.

Multimode step index (Fig. 6.4). This fibre has a relatively large core diameter which allows light to propagate at different angles or multimodes, and to have therefore, different transit times along its length. This undesirable effect is known as dispersion. If a pulse is transmitted it becomes stretched as it is propagated over several paths.

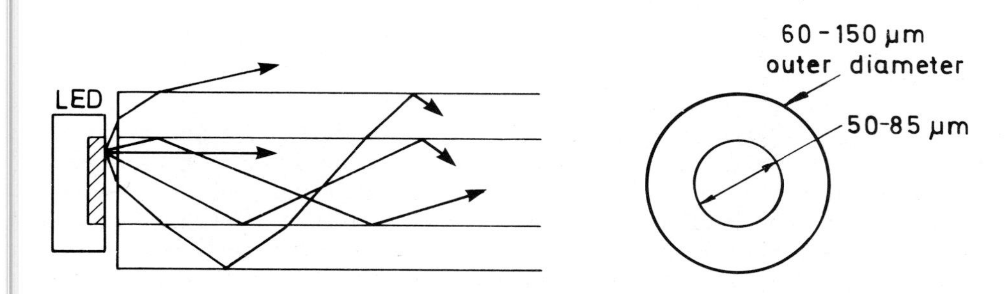

Fig. 6.4 Propagation along multimode step index fibre

Fibre specifications usually give bandwidth in units of MHz-km. For example a 200 MHz-km cable can send 200 MHz up to 1 km or 100 MHz up to 2 km, before the pulses stretch sufficiently to corrupt one another.

Monomode step index (Fig. 6.5). The core diameter is made sufficiently small so that multipath transmission is almost completely eliminated. However coupling to light sources and tolerances in connectors are more difficult than with multimode fibres. With these fibres, losses of fractions of dB's/km have been achieved.

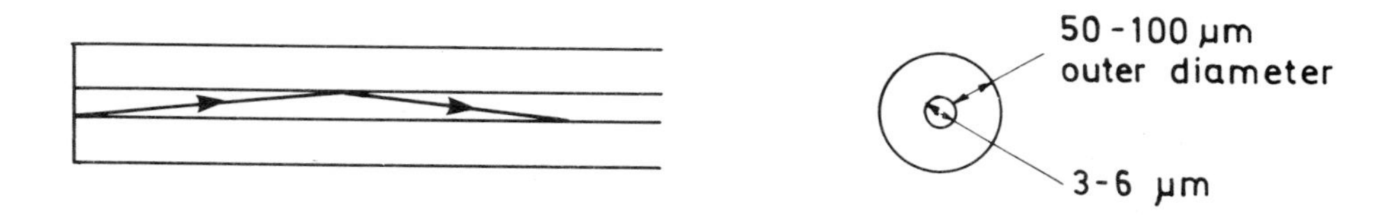

Fig. 6.5 Propagation along monomode step index fibre

Multimode graded index (Fig. 6.6). Instead of having a distinct refractive index difference, this fibre has a radially changing index. Light entering at different angles can thus be caused to travel at higher velocities away from the centre of the cable so that the same transit time is achieved for all paths. Losses of less than 1.5 dB/km have been realised.

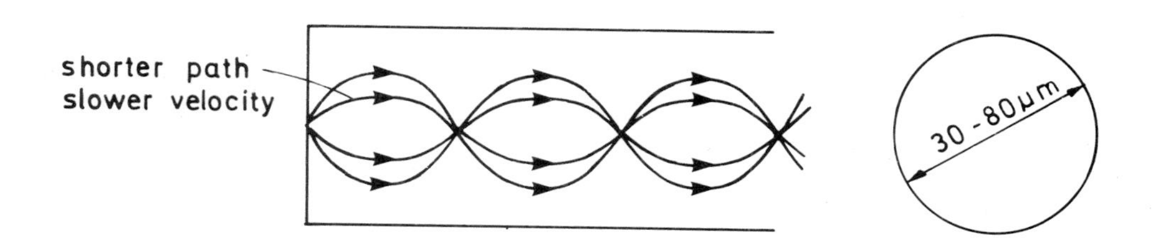

Fig. 6.6 Propagation along multimode graded index fibre

Power Budget

The light sources used are either light emitting diodes (LED) or lasers which give more power, and with careful matching several mW can be launched into the cable. The photo diodes which convert the light into an electrical signal, are either PIN or avalanche diodes which additionally amplify the signal. These components dictate the losses which can be tolerated in the fibre, including any joints. If the loss is excessive then repeaters must be included within the link. Providing electrical power for these repeaters may be difficult because fibre can convey very little power.

Military Factors

Most civil effort has gone into the production of low loss, low dispersion cable. This maximises the capacity and repeater spacing which are the two most important features of telephone trunk links. The majority of permanent military point-to-point links can use standard system designs. Typically 140 Mb/s systems have repeater spacings of 10-15 km using graded index fibre and in excess of 30 km with monomode fibre.

For most military purposes, capacity is not the limiting factor, instead operational constraints introduce different requirements. Although monomode fibre is smallest and lightest and has the best attenuation and bandwidth, the thicker fibre types can be more suitable. Factors which improve with thickness are the launch efficiency, connectability, performance over wide temperature ranges, effect of tight bends on both optical and mechanical properties, degradation caused by radiation induced attenuation (EMP), and response to frequent robust handling and reconnection.

SPEECH COMPRESSION

In military communications one of the main advantages of digitising speech is that encryption then becomes both simple and extremely secure. The two most common digitising techniques, PCM and delta modulation were described in Chapter 4. In both cases the bandwidth required to transmit the resulting pulse streams was considerably greater than that of the original analog speech. Thus these techniques can only be used for radio frequencies of VHF and above, where there is adequate bandwidth. For example with the CLANSMAN VHF net radio, the use of secure delta modulation rather than insecure analog speech results in a doubling of the required channel spacing. As this effectively halves the number of available frequency slots, frequency allocation, in an already crowded band, is more difficult. In the HF band conventional secure speech methods are completely impractical. What is required is a digitising method capable of providing satisfactory quality speech from a 2.4 kb/s stream rather than the 16 kb/s required for delta modulation.

It is known that the speech signal contains much redundancy. Thus there is a considerable difference between the rate at which the vocal mechanism appears to generate information and the minimum information rate required to communicate a message. If some of the redundancy is removed then the bandwidth or channel capacity can be reduced. To achieve secure HF communication on a 3 kHz channel, redundancies should be removed from the speech to reduce its bandwidth. Then the subsequent application of a conventional digitising process, such as PCM, will not increase the required bandwidth above 3 kHz.

Figure 6.7 shows the frequencies generated by a speaker in a typical sentence. Ignoring the rapid fluctuations it can be seen that the power, within small bands of frequencies, varies comparatively slowly and this property is made use of in vocoders. The most promising types are the channel vocoder and the linear predictive coder both of which can reduce the bit rate to 2.4 kb/s, which is a suitable rate for an HF channel.

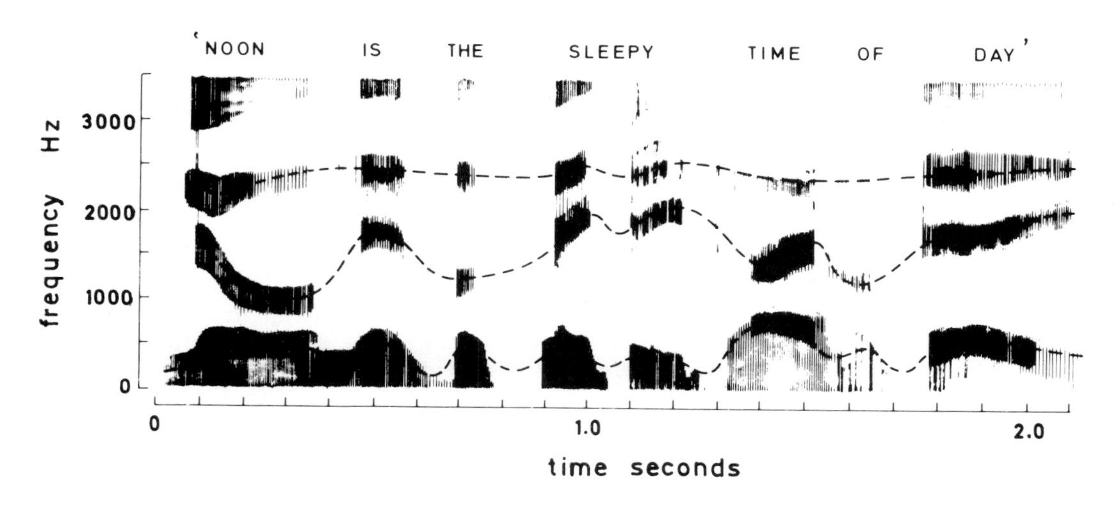

Fig. 6.7 Sound spectrogram

Channel Vocoder

In this device the speech input is divided into a number of bands, say 12, by a bank of band pass filters. The power in each band is measured and this produces a set of fairly low frequency signals. These signals together with an excitation signal containing information on the pitch and type of voicing, are transmitted and reassembled at the receiver where speech is reproduced.

Linear Predictive Coder

In this device the filter bank is replaced by a model of the vocal tract. Physically this can be regarded as a series of tubes of different diameters and is modelled electrically, by a series of filters. An analysis of speech segments provides a set of coefficients which vary these filters whilst the initial excitation signal is estimated in the same way as for the channel vocoder.

Alternative Speech Compression Techniques

Many alternative types of vocoder have been studied. In Fig. 6.7 it can be seen that the speech spectrum is made up from about three formants or resonant frequencies, and that these are slowly varying. Formant vocoders transmit these formants and the types of excitation. Voice-excited vocoders use the low frequency content of speech as an improved excitation signal. This improves the naturalness of the reformed speech but does not give such a large saving in bandwidth.

Time encoded speech is an alternative method using the time waveform rather than the frequency spectrum. The waveform is reduced to a set of numbers which describe the length or time of the waveform between crossings of a reference zero level and the number of minima or ripples within this length. At the receiver a reasonable approximation to the original waveform can be reconstructed. This method is quite simple and cheap to implement but does not offer as large a reduction in capacity as the vocoder.

Speech compression devices are capable of providing the desired low bit rate digital speech but their introduction has been quite slow. Examples of equipment fielded by the military are the German ELCROVOX which provides secure telephones within an insecure trunk system and BURGANET which gives secure speech over 2.4 kb/s links of the UK SKYNET satellite. Widespread introduction of vocoders requires a subjective decision of an acceptable standard, and comparisons between different vocoders usually reveal differences of a qualitative nature which make any choice difficult. The modern telephone line is the standard against which judgement should be made, it provides a reasonable amount of voice recognition and personality of the speech. It is often said that the speech quality of a communication system must be sufficient to allow the listener to gauge the personality of a stranger and the mood of a friend. While a commonly expressed army view is that it is essential for voice communications to allow commanders to give weight to their orders through their personalities and to judge a situation from the voice of a reporter as well as his words. A further problem in evaluating vocoders comes from the manner in which speech is degraded in the presence of noise. Typically, impulses of interference are heard as sharp sounds in analog systems but result in some slurring to words in vocoder systems. However it is likely that some reduction in speech quality is acceptable to the military in exchange for better spectrum utilisation and secure communications.

ADAPTIVE CANCELLATION

The aim of adaptive cancellation methods is to cancel out automatically, the effects of unwanted signals. These signals may be unintentional interference from colocated transmitters such as rebroadcast radios (EMC) or intentional jamming from remote transmitters (ECM).

In net radio rebroadcast, the co-sited transmitter and receiver use different frequencies and physically separated antennas. The degree of electrical and physical isolation must be sufficient for the filtering of the receiver to reject any signal which is picked up from the transmitter. The isolation can be improved by using an adaptive cancellation technique. The output of the receiver is connected to the transmitter input and a sample of the transmitted output signal is fed back to the receiver. The amplitude and phase of this sample is adjusted to cancel out exactly, any signal picked up due to the transmitter's radiation. Although there are technical difficulties in engineering this method, interference attenuation improvements of several tens of decibels have been achieved. A similar technique can be applied to jamming by using an additional receiving antenna to sample the interference which is then subtracted from the corrupted signal. This method has similarities with the sidelobe canceller form of null-steering antenna described in the ECCM technology section of Chapter 5.

It is now possible to make a single frequency rebroadcast station by using an adaptive cancelling technique to suppress the transmitted signal at the colocated receiver. The system effectively forms a narrow band or notch filter which tracks and removes the instantaneous frequency of the unwanted re-radiated FM signal. As there are delays both in the propagation path between the transmit and receive antennas and in the electronic circuitry, the received wanted and unwanted signals at any instant are at slightly different frequencies. Thus the notch only transiently effects the wanted signal. Installations have been demonstrated which can rebroadcast satisfactorily with power ratios of 130 dB between transmitted and received signals using the same carrier frequency.

SIGNAL PROCESSING

When a radio signal is received it is processed to extract its message content. For example if the signal is uncorrupted noise free frequency modulation, then the signal processing circuitry is simply an FM demodulator. More elaborate signal processing can optimise the extraction of signals embedded in noise or other interference. Most signal processing has been carried out by analog circuitry at the intermediate frequency of the receiver but alternative technology is now being used to give improved performance.

Surface Acoustic Wave (SAW) Devices

Surface acoustic wave devices are components that exploit the characteristics of a wave that propagates on the surface of a solid. Electrical coupling to the SAW is shown in Fig. 6.8. The transducer consists of a pattern of interleaved electrodes engraved in a thin metal film deposited on the surface of a piezo electric crystal substrate. When a radio frequency source is applied a wave is generated. A key feature of these devices is that the wave is accessible at any point along the length of the device.

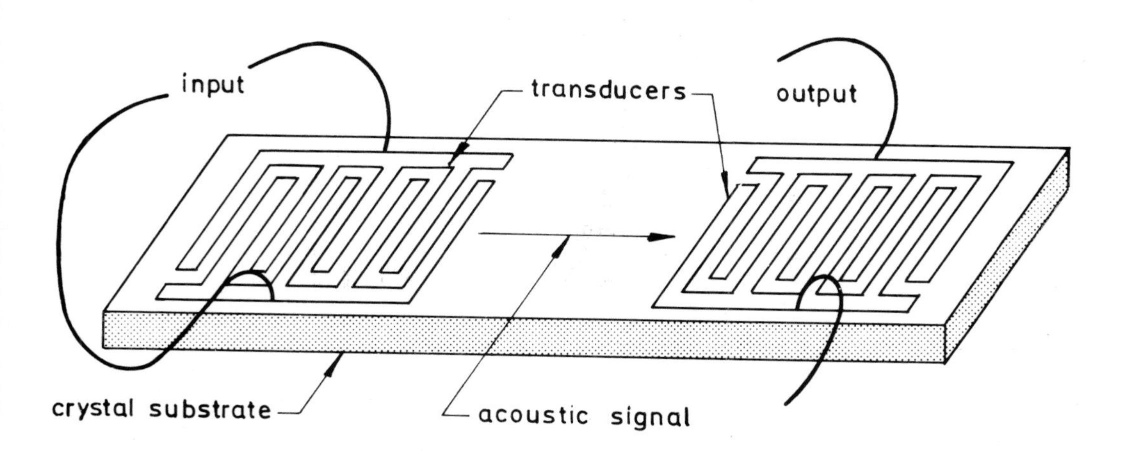

Fig. 6.8 Typical SAW device

The velocity of SAW propagation is typically 3000 m/sec whereas electromagnetic waves travel at 3 x 10^8 m/sec in free space. This leads to a spatial compression of 10^5, ie a 5µS signal occupying 1 km in free space is compressed to 1 cm on the surface of the SAW device. Most SAW devices constructed to date are used for frequencies between 30 and 1000 MHz which allows communications signal processing to be carried out directly at radio frequencies.

With SAW devices, because the signal can be accessed at any point on the surface it is possible, by sampling, modifying and recombining, to design a variety of signal processing devices. For example fixed and variable delays, fixed and swept frequency oscillators and band pass filters, can be readily formed. More elaborate structures can perform complex mathematical operations on the signals.

An important application is in spectrum analysis for ESM. A scanning receiver uses a narrow band filter that is swept slowly across the frequency band of interest. This method has high resolution and dynamic range but only examines a small part of the spectrum at any given time. The simple expedient of a bank of narrow band pass filters covering the range of frequencies of interest is the basis for the channelised receiver. SAW technology has revitalised this approach as SAW filters are compact, easily designed and replicated.

An alternative method for the measurement of frequency components can be achieved in microseconds by another SAW device. A single SAW component produces the Fourier Transform of a received waveform, this being the mathematical operation which directly produces the frequency spectrum. This technique, although fast, is limited in resolution (20 kHz) and dynamic range (50 dB).

Digital Signal Processing

The elements of a digital signal processor are shown in Fig. 6.9. The interface circuits are the analog-digital (A/D) and digital-analog (D/A) converters. The

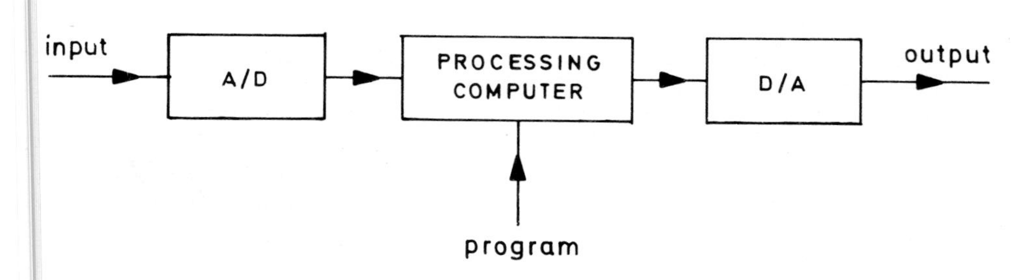

Fig. 6.9 Digital signal processor

rate of the bit stream produced by the A/D depends on the number of samples/second and the resolution of each sample (bits/sample). These respectively govern the bandwidth and dynamic range of the system. The processing computer must be able to carry out its operations on these bits at least as fast as they are generated. Special processors have been designed for communications signal processing because standard microprocessors are too slow. Currently it is possible to process input signals of some tens of kHz. The advantages of a digital signal processor are first of all, its programmability in that the same unit can perform alternative tasks by software: for example an IF filter can be changed for upper sideband SSB, lower sideband SSB, DSB or CW reception. Secondly, the processor has good stability and repeatability. Tolerance, ageing and component drift problems are largely eliminated.

SECTION II - SYSTEMS

STRATEGIC SYSTEMS

For communications, outside the combat zone, several communications systems are required to satisfy the complete range of military signalling needs. Long range systems can support a worldwide operational role and isolated long distance commitments. Local systems link static headquarters and other locations with military sole-user communications networks and provide links to the tactical systems.

Worldwide Communications

At one time communications systems of the army, navy and air force were maintained in parallel and largely independent of each other. Today combined tri-service systems are more common for long range communications and this is of great assistance to combined operations. Worldwide radio can be achieved by either HF skywave using ionospheric reflection or by satellites. HF provides relatively narrow bandwidth, variable quality links, less than twenty-four hour coverage and requires frequent frequency changes due to variations in the ionospheric layers. However for many years it has enabled small radios with simple antennas to communicate around the world.

Satellite systems are continually improving and, provided there is a suitable satellite in position, high quality twenty-four hour communications are possible. The distance between the ground stations is unimportant, although they must be within the surface area illuminated by the satellite. Secure speech and telegraph can be provided and communications are highly resistant to enemy intercept and DF. The satellites orbit at high altitudes and are difficult to destroy but the initial cost to establish the space segment is high and there have been many unsuccessful launches. An additional weakness is that faults in the satellite electronics and position control are almost irrepairable.

Beyond the Horizon Communications

This is another area in which many HF links have been replaced. Satellite links are again an obvious alternative but in addition tropospheric scatter systems are suitable for links of up to several hundred kilometres. Two typical examples of static tropospheric scatter links were shown in Table 2.1 of Chapter 2.

Many secure, high quality, reliable channels are provided and, because directional antennas are used, enemy ECM is made difficult. Mobile tropospheric scatter links also exist but are less common because they require large power supplies and large antennas. As the antennas must be accurately aligned in unobstructed sites, their concealment is difficult and they present a radiation hazard.

Radio relay is also used for static strategic links. Unlike the tactical mobile links, microwave frequencies are preferred due to the extra bandwidth available. The preference for lower frequencies in mobile links is because the resulting lower gain antennas are easier to align.

Local Networks

Although most well developed countries have a comprehensive structure of trunk communications the military often provide themselves with an extensive network of sole-user communications for a variety of reasons. Civil circuits are vulnerable to industrial action and sabotage and as they are based on population centres, communications are liable to damage in general war. In addition the nature and location of defence requirements renders the civil solution an expensive one in terms of rental. Civil communications are relatively inflexible and the current quality is inhibited by the large investment in analog equipment.

In broad terms the characteristics of the military systems which replace the civilian networks should overcome these drawbacks. They should be guarded, if not manned by military personnel and should be based on militarily defensible areas away from major population centres. A sufficient proportion of all military traffic should be carried in peace time so as to be cost effective when compared with civilian rental charges. The systems make full use of digital and other techniques to provide high quality circuits with an element of flexibility, perhaps involving some mobile equipment. Reliability, storage considerations, resistance to EMP, test and repair criteria should conform to military standards and interoperability with other strategic and tactical systems should be more easily achieved.

Interoperability

Although it is relatively simple to set up a purpose built communication system for every requirement, it can lead to there being an excessively large number of independent systems which may be impractical to interconnect. Such an approach is expensive and inflexible in reacting to change. Some thought about interoperability can minimise these problems. Interoperability is not a state which exists or does not exist, rather there are various levels which depend on the technical interface possibilities, and equally importantly, on the management and control philosophies applicable to the systems concerned. The range of technical interfaces varies from impractical to common equipment, as shown in Table 6.1. In the intermediate cases gateways or suitable interfaces, are established. If the systems are very dissimilar then the gateways have to carry out many functions and are expensive so that few are deployed. However with greater system standardisation, costs fall and the number of gateways can be increased to give multiple entry points, greater interoperability and enhanced survivability. Table 6.1 also shows the management and control philosophies which affect the degree of interoperability. Care must be taken if the interconnected systems have different levels of security or if the data bases of one system are not to be fully shared with users of the other.

TABLE 6.1 Levels of Interoperability

<table>
<tr><th>TECHNICAL INTERFACE</th><th>MANAGEMENT STATE</th><th>LEVEL OF INTEROPERABILITY</th></tr>
<tr><td rowspan="2">impracticable to interface systems</td><td>complete independence</td><td>Separate Systems</td></tr>
<tr><td>memorandum of understanding to share resources ie access to both systems</td><td>Shared Resources</td></tr>
<tr><td rowspan="2">feasible to develop interface box to allow interconnection</td><td>agreement for users to talk to one another with no impact on individual system design</td><td>Gateways</td></tr>
<tr><td rowspan="2">agreement for users to talk to one another with intersystem flexibility (eg survivability and overload consideration) but retain individual prerogatives</td><td>Multi-entry Gateways</td></tr>
<tr><td rowspan="2">one or both systems can be redesigned to allow connection between systems without using interface box</td><td>Compatible Systems</td></tr>
<tr><td>willingness to accept significant impact on system from actions taken by users and management of external system(s)</td><td>Completely Interoperable</td></tr>
<tr><td>both systems use common equipment - there are no technical interface problems</td><td>separate systems placed under common management/control of external system</td><td>Same System</td></tr>
</table>

There are many factors acting for and against interoperability, particularly in communications between foreign forces. Two important considerations are the effects of political alignment and of emerging technology.

Defence alignments normally occur within alliances such as NATO or the Warsaw Pact. Within such organisations interoperability of communications systems greatly assists in the liaison and control of operations across military or even national boundaries.

New technology as it emerges can cause interoperability problems. Communications can be improved by the use of new, more capable equipment. However its introduction is often restrained by the need to remain interoperable with existing systems. The result is that the procurement of military equipment seldom makes full use of the advantages of the latest technology.

TACTICAL SYSTEMS

The main tactical communication systems are net radio and trunk radio. Net radio was described in detail in Chapter 4; it provides all informed networks for mobile communication within units and formation command nets as far back as corps headquarters. Trunk radio communication was described in Chapter 5; it enables calls to be made to individuals by linking headquarters together and it operates at formation level with some extension rearward. An additional useful form of communication is to be able to selectively call, not only static subscribers but also mobiles. Such a feature has been added to some modern trunk systems.

Mobile Subscriber Access

Mobile or isolated users can be provided with the same service features as a static subscriber of a trunk system. In the British PTARMIGAN system this is achieved through the Single-Channel Radio Access (SCRA) sub-system. SCRA provides a facility similar to civilian radio-telephones used by subscribers travelling in their cars. By using SCRA a commander with a mobile terminal can make telephone calls to his own headquarters or any other mobile or static subscriber in the trunk system.

The complete area is covered by continuously transmitting SCRA radio centrals through which a mobile communicates. As the subscriber moves he is automatically reaffiliated to appropriate centrals in turn, as shown in Fig. 6.10. Mobiles receive signals from a central and affiliation is confirmed to the user by a light on his set. When the central has a call for a mobile, his number and the channels he should use are included in the affiliation signal. The SCRA mobile terminal automatically recognises its number and when the handset is lifted communicates with the central on the correct channels. Another automatic feature is that the mobile's transmitter power is adjusted to be adequate, but not excessive for the radio path used, thus minimising interference with other users of the VHF band. The user is not required to have any knowledge of these radio features and can use his telephone in the same way as a static subscriber.

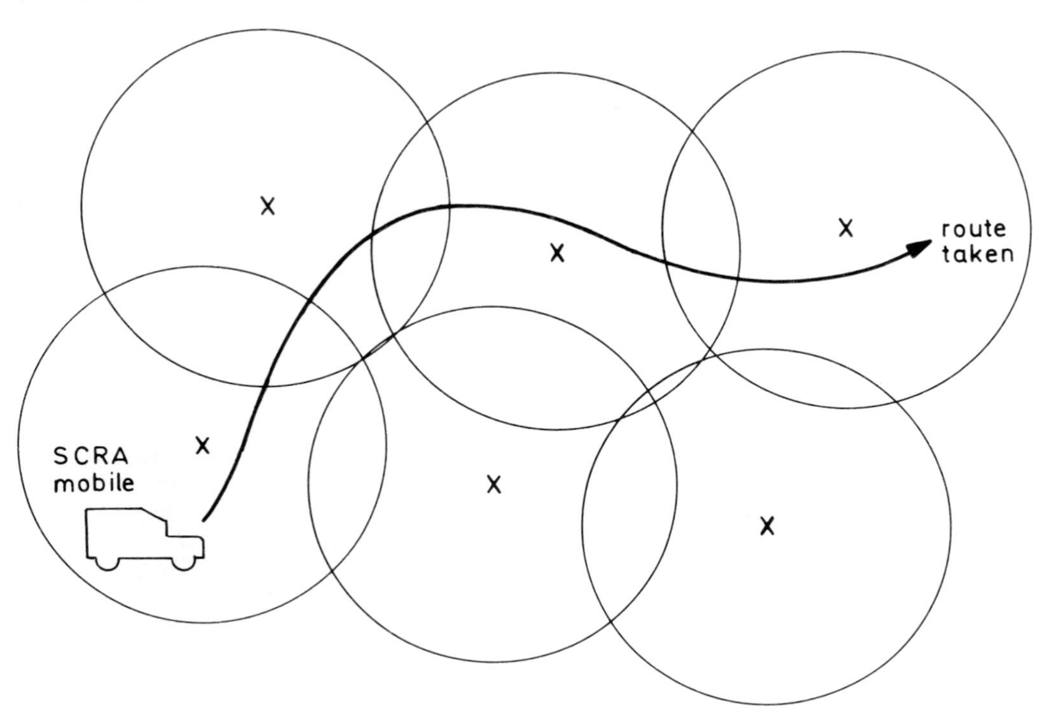

Fig. 6.10 Coverage of SCRA

Figure 6.11 shows an SCRA mobile subscriber terminal mounted in the rear of a Land Rover. The push button dialling and status lights can be seen on the upper box, the Control Indicator Group. On the right of the lower shelf is the SCRA radio. Also shown in this installation is a Combat Net Radio Interface (CNRI) in the centre of the lower shelf. This enables the combat net radio, on the left (a CLANSMAN VRC 353) to obtain access to the trunk system. By this means any net subscriber can take part in voice calls with any PTARMIGAN subscriber, and vice versa, using net radio procedures. This facility is available for both secure and insecure net radio users although care must be taken to maintain security when talking from a secure trunk network into an insecure radio net. Net subscribers may also participate in conference and broadcast calls.

The SCRA sub-system relies on the trunk system nodes for switching but if the routing control was moved to the central it would then be possible to have a 'Stand-Alone' SCRA system. Such a system could serve as a small trunk network or as the equivalent of a selective calling net.

Fig. 6.11 SCRA mobile terminal

Developments in Tactical Communications

The growth of and dependence on tactical communications for today's command and control requirements are creating two major problem areas. The available frequency spectrum is becoming increasingly congested and the potential damage successful enemy EW can cause is ever increasing. Modern developments are addressing both problems although in some cases improvement of one is at the expense of the other.

A modern fully deployed trunk network provides many redundant, secure links over directional UHF paths. It thus offers considerable resistance to enemy EW. The use of low powers and terrain screening gives a low probability of intercept and additionally enables frequencies to be re-used within the network's area. Higher UHF frequencies are now being used so that, in general, sufficient frequencies are available. A possible future enhancement would be the use of satellites to replace radio relay over some paths, thus making intercept even harder and giving more flexibility in the UHF frequency planning. Connections from headquarters to radio relay still pose problems due to long setting up times but a greater use of short SHF radio links and fibre optic lines may improve this situation.

SCRA type systems provide selective calling between mobiles and other subscribers. For some users selective calling may be a more appropriate method than continuously monitoring an all-informed net. This may be so for net radio users concerned with administrative support of the battle who only need to react when it directly concerns them. For this reason the introduction of SCRA might permit an equivalent reduction in the numbers of radio nets. This would certainly be welcomed from the frequency allocation viewpoint because SCRA and the majority of radio nets use the same, crowded, VHF frequency bands. However SCRA is still in its infancy and it would be a profound change for many users to lose the all-informed net as their primary means of communication.

In net radio, until a satisfactory standard for low bit rate, narrow band, digital speech is achieved, secure nets will be restricted to VHF. Even in this band a lower bit rate would be advantageous as spectrum congestion can be acute. At VHF, frequency modulation is usual but considerable attention is being given to narrower band alternatives and indeed PTARMIGAN SCRA already uses such a method. In general narrower band methods require greater stability from the local oscillator synthesisers and more linear amplifiers. Both of these factors lead to heavier equipment and greater power consumption. However it is quite possible that SSB will become common at VHF in the near future. An additional method of easing spectrum congestion is to use single frequency rebroadcast, the technique of which was described in the section on adaptive cancellation. On the other hand frequency hopping radios are being produced commercially and their widespread introduction would make frequency planning more complicated. Whether frequencies should be allocated in blocks or whether nets should use the same frequencies by synchronisation, is still unclear and there is little practical experience of how many interleaved or repeated frequency hopping nets would perform in relation to each other. The alternative of replacing only a few isolated, conventional nets by frequency hopping radios could well encourage the enemy to attack these specifically.

Traditionally communications systems have been designed to provide a certain number of speech channels between users. However, distributed ADP facilities are increasingly being provided in support of Command and Control staff. Thus the primary role of the communications system is becoming the coordination of ADP facilities by allowing data to flow securely and reliably between the subscribers, as well as the more conventional but less efficient, voice traffic. The increased use of data entry, burst transmission devices and the provision of channels of varying capacities in the trunk network are logical steps. The future will surely see further changes to increase the efficiency and speed of the passage of information and improve still further the facilities provided to achieve coordinated command and control.

SELF TEST QUESTIONS

QUESTION 1 Estimate the propagation loss of a link from a geostationary satellite to a ground station if the frequency used is 6 GHz.

Answer ....

....

QUESTION 2 How is it possible for many users to share the same satellite?

Answer ....

....

QUESTION 3 Where may optical fibres be used in military communications?

Answer ....

....

QUESTION 4 Why is there continuing interest in speech compression?

Answer ....

....

QUESTION 5 Discuss the ways in which modern signal processing techniques can enhance military radio equipment.

Answer ....

....

....

QUESTION 6 What factors should be considered when establishing flank communications?

Answer ....

....

QUESTION 7 What roles can satellite communications play now that small ground terminals are becoming available?

Answer ....

....

QUESTION 8 What are the implications of using an SCRA type system rather than a conventional radio net?

Answer ....

....

....

ANSWERS ON PAGE 138

Answers to Self Test Questions

CHAPTER 1

Page 6

QUESTION 1

a. Range of communications is stretched considerably and links become less reliable. Later chapters will show that this creates siting, EW and rebroadcast problems.
b. Increased traffic flow because level of command acts as a filter in reducing the traffic flowing back to higher levels.

QUESTION 2

In addition to the obvious communications needed for voice and data links, communications are required to both gather information from, and to control other sensors, eg radar, acoustic and other unattended ground surveillance devices, remotely piloted vehicles, etc.

QUESTION 3

a. C^2I is becoming increasingly reliant on more radio communications.
b. Increased use is being made of the electromagnetic spectrum by other users, eg radar, surveillance, aircraft, etc.

CHAPTER 2

Page 30

QUESTION 1

Any complex waveform can be expressed as a sum of simultaneous sinewaves. Spread of frequencies is the bandwidth of the signal waveform.

QUESTION 2

a. Natural and galactic noise.
b. Man-made noise.
c. Thermal noise.

QUESTION 3

a. Requires only half bandwidth of normal AM (DSB).
b. Carrier is suppressed, thus saving power.

QUESTION 4 Although more bandwidth is used, interfering signals are suppressed. (In Chapter 3 other 'capture effect' advantages are described.)

QUESTION 5 $C = W \log_2 (1 + P/N) = 16{,}000 \times \log_2 (1 + 7)$

$= 16{,}000 \times 3 = 48$ kbits/sec.

QUESTION 6 Ground plane is metal case of set but is small and may be raised from ground. In addition antenna may be poorly sited and not vertical.

QUESTION 7 Directional antennas such as:

a. Log-periodic.
b. Stacked dipole.
c. Corner reflector.

QUESTION 8 $\lambda = \frac{c}{f} = \frac{3 \times 10^8}{50 \times 10^6} = 6\text{m}$

$FSL = (\frac{4\pi r}{\lambda})^2 = (\frac{4\ \pi 10 \times 10^3}{6})^2$

$= 4.4 \times 10^8$ or 86 dB.

QUESTION 9
a. Surface wave.
b. Ionospheric reflection.
c. Tropospheric scatter.

QUESTION 10 Variations in the propagation path. Destructive interference between alternative paths.

CHAPTER 3

Page 45

QUESTION 1 Tactical levels and mobile operations by formations and units below corps headquarters level, where immediate use and fast response are important.

QUESTION 2
a. Noise should be low for good sensitivity.
b. Filtering combined with IF gives correct selectivity.
c. Linear amplification prevents distortion and minimises spurious frequency responses.

QUESTION 3 Transceivers drift independently from a very high nominal carrier frequency, if drift is excessive then distortion and channel overlap may occur.

QUESTION 4 HF

a. Narrow bandwidth.
b. Greater range, particularly using skywave.
c. Rebro's not required.
d. Small shadow effects.

VHF

a. Secure.
b. Shorter resonant antennas.
c. Greater spectrum for users.
d. Noise suppression due to capture effect.
e. Defined service areas due to capture effect.
f. Terrain screening can counter ESM (see Chapter 5).

QUESTION 5 Advantages

a. Cheap.
b. Quickly into service.
c. Small size.

Disadvantages

a. Usually two-frequency simplex which is relatively unfamiliar to military users.
b. Lack of robustness.
c. Do not satisfy military temperature and environment specifications.
d. Ease of maintenance, not in initial design.
e. Not nuclear hardened.

QUESTION 6 Frequency planning reduces problems of interference. At VHF, frequencies can be re-used elsewhere on the battlefield.

CHAPTER 4

Page 77

QUESTION 1 a. Chain of command system control is responsibility of the staff of HQ concerned. In an area system overall deployment affects location of relevant signal staff.
b. Clearly an area system is more flexible and communication is always possible between any pair of subscribers.
c. Chain system can be broken whilst in an area system alternative routes bypass damage.

d. In area system the capture of a subscriber gives enemy access to complete system, less so in hierarchical system. Measures taken may include frequent changes of codes, destruction of equipment to avoid compromise and visits to isolated detachments.

QUESTION 2

Telegraphy:

Slow method but economical on bandwidth (several/channel). Teleprinter and skilled operator required. Good for multiple address and broadcast messages. Special store and forward equipment can give delayed delivery.

Telephony:

Best method for conversation and conveying features of speaker's voice. Only one user at a time on a standard communications channel.

Facsimile:

Good for maps and original documents. No retyping is required. Usually uses a complete channel. Modern equipment speed is approximately 1 minute for A4 document.

Television:

Requires considerable bandwidth. Only useful in specialised roles such as surveillance and remote briefings.

QUESTION 3

Advantages

a. Ease of encryption.
b. Periodic regeneration maintains quality.
c. Can use same links as data.
d. Consistent with modern electronic technology.

Disadvantages

a. Greater bandwidth required.

QUESTION 4

Telephony:

a. Analog electromechanical, eg Strowger, crosspoint, reed relay.
b. Digital crosspoints, eg PTARMIGAN.

Telegraphy:

a. Through switching, as telephony.
b. Tape relay.
c. Store and forward.

QUESTION 5 Facilities of PTARMIGAN are listed in Chapter 4. Software control permits their modification and allows new facilities to be added.

CHAPTER 5

Page 106

QUESTION 1
a. Good dynamic range and sensitivity.
b. Rapid sweeping of band or channelised approach.
c. Panoramic display and automatic logging.
d. Adequate frequency resolution.

QUESTION 2
a. Tactical information, eg HQ locations, net activity.
b. Steerage information for offensive EW, ie jamming and deception.

QUESTION 3
a. For accurate DF, stations located on a base line.
b. Site should be open and well forward.
c. Tactically some concealment is essential.
d. Minimise likelihood of false bearings by avoiding obstructions.

QUESTION 4 Advantages:

a. Can jam enemy communications with minimum effects on friendly communications.
b. Ability to win 'power battle' by being close to target.
c. Can be deployed in large numbers by limited manpower.
d. Enemy requires substantial resources to locate and destroy.
e. Cheap.
f. 'Smart' form can be triggered when required.

Disadvantages:

a. Problems with delivery systems.
b. Limited life and transmitter power due to small power source.
c. Small antennas are inefficient and radiation patterns may not give complete coverage.

QUESTION 5
a. Minimise time and power of transmissions, ie good EMCON policy.
b. Change frequency or location, or use alternative forms of communication.
c. Continue to transmit as before whilst using alternative frequency or forms of communication.
d. Authentication, standard operational procedures and voice recognition.

QUESTION 6

a. Technical problems now largely overcome, but reliable synchronisation is essential.
b. Spectral pollution is aggravated. Fast hopping rates are desirable to defeat fast-follower-jammers but pollution is increased further.
c. Compatibility with existing radios and nets.
d. Frequency planning; either in blocks or by synchronisation between different nets.
e. Many nets must be deployed to give anonymity.

QUESTION 7

Techniques are listed in Table 5.1. All except DSSS could be used in net radio. No one method should be relied on, the ability to call on several approaches is advantageous.

CHAPTER 6

Page 130

QUESTION 1

$$\lambda = \frac{c}{f} = \frac{3 \times 10^8}{6 \times 10^9} = 0.05 \text{ metre.}$$

$$\text{FSL} = \left(\frac{4\pi r}{\lambda}\right)^2 \text{ from Chapter 2.}$$

$$= \left(\frac{4\pi\ 36{,}000 \times 10^3}{0.05}\right)^2 = 8.2 \times 10^{19} \text{ or } 199 \text{ dB.}$$

In practice atmospheric absorption adds a few dB's.

QUESTION 2

Each user has his own frequency, time slots or code. TDM or FDM multiplexed channels are handled by the satellite. Multiple access techniques allow terminals, even when geographically dispersed, to use these channels.

QUESTION 3

As replacements for existing cables to radio relay, nets and around headquarters. They provide greater capacity, reduced weight, good EMP and EMC performance but connector and splicing problems remain.

QUESTION 4

If satisfactory quality speech can be obtained from a 2.4 kb/s digital stream, then conventional channels including HF net radio, can be made secure. Similar compression methods at VHF would provide more channels in the same spectrum.

QUESTION 5

a. Provide extra facilities, better displays and simplify operation (user friendly).
b. Greater reliability and repeatability from digital signal processors.

c. Versatility from software control of processing.
d. Extraction of more information from radio signals, eg ESM receivers.

QUESTION 6 Level of interoperability depends on technical interface possibilities and management state. Political considerations apply especially in terms of security and freedom of access to information in data bases.

QUESTION 7
a. Long distance, high capacity, secure strategic links with twenty-four hour availability.
b. Short distance, secure tactical links between manpack terminals with good ESM and ECM characteristics.

QUESTION 8
a. Conventional nets are all informed and require continuous monitoring. SCRA is equivalent to a selective called net.
b. SCRA centrals transmit continuously; nets are silent except when there is traffic.
c. One SCRA could replace many underused nets, but nets are most appropriate for well forward positions.

Glossary of Abbreviations

A

A/D - Analog-Digital
ADP - Automatic Data Processing
AGC - Automatic Gain Control
AM - Amplitude Modulation
ATE - Automatic Test Equipment
ATU - Antenna Tuning Unit

B

bit - binary digit
BITE - Built-In Test Equipment
b/s - bits/second

C

C^3 - Command, Control and Communications
C^2I - Command, Control and Intelligence
CDMA - Code Division Multiple Access
CNRI - Combat Net Radio Interface
COMCEN - Communications Centre
COMHD - Communication Head
COMINT - Communications Intelligence
CRO - Cathode Ray Oscilloscope
CW - Continuous Wave

D

D/A - Digital-Analog
dB - decibel
dBW - dB relative to IW
DF - Direction Finding
DSB - Double Side Band
DSSS - Direct Sequence Spread Spectrum

E

ECM - Electronic Counter Measures
ECCM - Electronic Counter Counter Measures
EDC - Error Detecting Code
EMC - Electro Magnetic Compatability
EMCON - Emission Control
EMP - Electromagnetic Pulse
ESM - Electronic Support Measures
EW - Electronic Warfare

F

f_c - Critical Frequency
FCL - Frequently Called List
FDM - Frequency Division Multiplexing
FDMA - Frequency Division Multiple Access
FH - Frequency Hopping
FM - Frequency Modulation
f_{muf} - Maximum Usable Frequency
FSL - Free Space Loss

G

G - Gain
GHz - Gigahertz (10^9 Hz)
G/T - Gain/Temperature

H

HF - High Frequency
HQ - Headquarters
Hz - Hertz

I

IC - Integrated Circuit
IF - Intermediate Frequency

K

kb/s - kilobits/sec (10^3b/s)
kHz - kilohertz (10^3Hz)
km - kilometre (10^3m)

L

LED - Light Emitting Diode
LO - Local Oscillator
LPI - Low Probability of Intercept

M

m - Modulation index
MEROD - Message Entry and Read Out Device
Mb/s - Megabits/sec (10^6b/s)
MHz - Megahertz (10^6Hz)
MUF - Maximum Usable Frequency
mW - milliwatt (10^{-3}W)

P

P - Power
PCM - Pulse Code Modulation

R

RF - Radio Frequency
Rx - Receiver

S

SAW - Surface Acoustic Wave
SCRA - Single Channel Radio Access
SDMA - Space Division Multiple Access
SHF - Super High Frequency
S/N - Signal/Noise
SSB - Single Side Band

T

TARIF - Telegraph Automatic Routing In the Field
TDM - Time Division Multiplexing
TDMA - Time Division Multiple Access
TPL - Transmission Path Loss
Tx - Transmitter

U

UHF - Ultra High Frequency

V

VFT - Voice Frequency Telegraphy
VHF - Very High Frequency

W

W - Watt

Index